Lecture Notes in Computer Science 11963

More information about this series at http://www.springer.com/series/7412

Yunliang Cai · Liansheng Wang ·
Michel Audette · Guoyan Zheng ·
Shuo Li (Eds.)

Computational Methods and Clinical Applications for Spine Imaging

6th International Workshop and Challenge, CSI 2019
Shenzhen, China, October 17, 2019
Proceedings

 Springer

Editors
Yunliang Cai
Worcester Polytechnic Institute
Worcester, MA, USA

Michel Audette
Old Dominion University
Norfolk, VA, USA

Shuo Li
Western University
London, ON, Canada

Liansheng Wang
Xiamen University
Xiamen, China

Guoyan Zheng
Shanghai Jiao Tong University
Shanghai, China

ISSN 0302-9743   ISSN 1611-3349   (electronic)
Lecture Notes in Computer Science
ISBN 978-3-030-39751-7   ISBN 978-3-030-39752-4   (eBook)
https://doi.org/10.1007/978-3-030-39752-4

LNCS Sublibrary: SL6 – Image Processing, Computer Vision, Pattern Recognition, and Graphics

This Springer imprint is published by the registered company Springer Nature Switzerland AG
The registered company address is: Gewerbestrasse 11, 6330 Cham, Switzerland

Preface

The spine represents both a vital central axis for the musculoskeletal system and a flexible protective shell surrounding the most important neural pathway in the body, the spinal cord. Spine related diseases or conditions are common and cause a huge burden of morbidity and cost to society. Spine imaging is an essential tool for assessing spinal pathologies. Giving the increasing volume of imaging examinations and the complexity of their assessment, there is pressing need for advanced computerized methods that support the physician in diagnosis, therapy planning, and interventional guidance.

The objective of this combined workshop and challenge on spinal imaging was to bring together researchers who share a common interest in spine focused research and to attract additional researchers to this field. By allowing both submissions of papers on novel methodology and clinical research, and also papers which demonstrate the performance of methods on the provided challenges, the aim was to cover both theoretical and very practical aspects of computerized spinal imaging.

We invited spine imaging researchers to share and exchange their experiences and expertise in spinal imaging and method development. Prof. Shisheng, Professor of School of Medicine from Tongji University, Shanghai, China, who is also the chief physician and chief director of the spinal surgery, Department of Orthopedic, Shanghai Tenth People's Hospital, Shanghai, China, gave the keynote speech. He shared his experience about the clinical challenge and opportunity of deep learning techniques from a surgeon's perspective. The talk, with a full house audience, attracted not only all CSI participants but also many other MICCAI attendantes on the workshop day.

The increasing number of publications in recent years on spinal imaging, in particular at MICCAI, indicate the high relevance of this topic to the community. After five successful workshops at MICCAI 2013, 2014, 2015, 2016, and 2018, we also had an increased number of participants in this year's workshop. The workshop selected five regular papers on spine image analysis, covering the topics of vertebra detection, spine segmentation, and image-based diagnosis. Each submission was rigorously reviewed by two or three Program Committee (PC) members on the basis of its technical quality, relevance, significance, and clarity. The Best Paper Award was given to the paper "Conditioned Variational Auto-Encoder for Detecting Osteoporotic Vertebral Fractures" by Malek El Husseini, Anjany Sekuboyina, Amirhossein Bayat, Bjoern Menze, Maximilian Loeffler, and Jan Kirschke, based on the raw scores of all review feedbacks.

In addition to regular research presentations, the computational challenge was organized to attract researchers working on general purpose algorithms to try their methods on spinal data. The MICCAI Challenge on Accurate Automated Spinal Curvature Estimation (AASCE) 2019 was jointly organized with the CSI 2019 workshop. The goal of the challenge was to investigate (semi-)automatic spinal curvature estimation algorithms and provide a standard evaluation framework with a set of x-ray images. The challenge attracted 79 international participating teams. The top eight teams were invited for presentation in the workshop. The Tencent YouTu Lab team achieved the best performance on all metrics. The short papers of the AASCE 2019 challenge participants were included in the workshop post-proceedings. These short papers focused on presenting the methodologies used for the challenge task.

We would like to thank the MICCAI workshop organizers for supporting the organization of the CSI 2019 workshop, as well as all of the PC members for their great efforts and cooperation in reviewing and selecting the papers. We would also like to thank all of the participants for attending the regular presentation sessions and challenge competition session. Finally, our gratitude goes to Springer, especially Anna Kramer, for their continuous support in the publication of the workshop proceedings.

November 2019

Yunliang Cai
Liansheng Wang
Michel Audette
Shuo Li
Guoyan Zheng

Accurate Automated Spinal Curvature
Estimation Challenge 2019

1 Introduction

Scoliosis is often defined as a curvature of the spine. The Cobb angle is the most widespread method used for determining the degree of scoliosis. Cobb angle greater than 45 degrees is considered severe scoliosis. Only a few methods have been proposed for the automatic estimation of Cobb angles. The purpose of the accurate automated spinal curvature estimation (AASCE) challenge is to provide an objective evaluation platform for automatic spinal curvature estimation algorithms.

2 Task

The task of this challenge is to automatically obtain 3 Cobb angles from each spinal anterior-posterior X-Ray image. Participants need to submit a CSV file of the Cobb angles results of the test set. We use an online automatic evaluation system (https://aasce19.grand-challenge.org).

Dataset The dataset contains a total of 707 spinal anterior-posterior x-ray images. Among them, 609 are the training set, and 98 are the testing set. In the open training set, the landmarks were provided, and the Cobb angles were calculated using these landmarks. The testing set only released x-ray images with the ground truth only known by organizers. These annotations were done manually by professional doctors.

Evaluation Submissions are scored based on a symmetric mean absolute percentage (SMAPE). The SMAPE metric is defined as:

$$SMAPE = \frac{1}{N} \sum_{i=1}^{N} \frac{SUM|X_i - Y_i|}{SUM(X_i + Y_i)} \times 100\%$$

where X_i is the estimated Cobb angles, Y_i is the ground truth, N is the number images.

Results Table 1 depicts the 5 top-ranked results of the AASCE-2019, after cleaning up some false and duplicate submissions. Team X won this task with 21.7135% SMAPE.

Table 1. Mean performance and ranking of the 5 top-ranked team on SMAPE.

Ranking (#)	Team	SMAPE (%)
1	X	21.7135
2	iFLYTEK	22.1658
3	Erasmus MC	22.9631
4	vipsl	24.7987
5	JLD	25.4784

3 Conclusion

In this paper, we provide an overview of the setup and results of the AASCE-2019 challenge at the 2019 MICCAI Workshop on Computational Methods and Clinical Applications for Spine Imaging. As further work, the challenge will be reopened for a new submission.

December 2019

Organization

General Chairs

Yunliang Cai   Worcester Polytechnic Institute, USA
Liansheng Wang Xiamen University, China
Michel Audette   Old Dominion University, USA
Shuo Li   Western University, Canada
Guoyan Zheng   Shanghai Jiao Tong University, China

Contents

Regular Papers

Detection of Vertebral Fractures in CT Using 3D Convolutional Neural Networks

Joeri Nicolaes[1,2,3](✉) [ID], Steven Raeymaeckers[4], David Robben[2,3,5] [ID],
Guido Wilms[6] [ID], Dirk Vandermeulen[2,3] [ID], Cesar Libanati[1], and Marc Debois[1]

[1] UCB Pharma, Brussels, Belgium
joeri.nicolaes@ucb.com
[2] Medical Imaging Research Center (MIRC), KU Leuven, Leuven, Belgium
[3] Medical Image Computing (MIC), ESAT-PSI,
Department of Electrical Engineering, KU Leuven, Leuven, Belgium
[4] Department of Radiology, University Hospital, Brussels, Belgium
[5] icometrix, Leuven, Belgium
[6] Department of Radiology, UZ Leuven, Leuven, Belgium

Abstract. Osteoporosis induced fractures occur worldwide about every 3 s. Vertebral compression fractures are early signs of the disease and considered risk predictors for secondary osteoporotic fractures. We present a detection method to opportunistically screen spine-containing CT images for the presence of these vertebral fractures. Inspired by radiology practice, existing methods are based on 2D and 2.5D features but we present, to the best of our knowledge, the first method for detecting vertebral fractures in CT using automatically learned 3D feature maps. The presented method explicitly localizes these fractures allowing radiologists to interpret its results. We train a voxel-classification 3D Convolutional Neural Network (CNN) with a training database of 90 cases that has been semi-automatically generated using radiologist readings that are readily available in clinical practice. Our 3D method produces an Area Under the Curve (AUC) of 95% for patient-level fracture detection and an AUC of 93% for vertebra-level fracture detection in a five-fold cross-validation experiment.

Keywords: Vertebral fracture · 3D Convolutional Neural Networks · Detection

1 Introduction

Current radiology practice grades vertebral fractures according to Genant's semi-quantitative Vertebral Fracture Assessment (VFA) method [5]. This method assesses the vertebral body morphology in X-ray images or at/around the mid-sagittal plane in 3D image modalities (CT, MR). As reported by Buckens et al. [3], the intra- and inter-observer reliability and agreement of semi-quantitative VFA on chest CT is far from trivial on patient- and vertebra-level.

© Springer Nature Switzerland AG 2020
Y. Cai et al. (Eds.): CSI 2019, LNCS 11963, pp. 3–14, 2020.
https://doi.org/10.1007/978-3-030-39752-4_1

A number of publications on vertebral fracture detection are inspired by how radiologists apply the Genant classification: firstly they attempt to segment the vertebrae at high accuracy, secondly the endplates are detected and finally the height loss of each vertebra is quantified in order to detect vertebral fractures. Such methods rely exclusively on 2D [2] and 2.5D [14] height features.

Valentinitsch et al. propose a pipeline that first segments the vertebrae, then extract various 3D texture features (e.g. Histogram of Oriented Gradients, ...) and volumetric Bone Mineral Density (vBMD) to finally apply a Random Forest classifier for patient-level fracture detection. Their experimental results show that combining multiple features calculated for each vertebra along the spine yields superior results [12]. Bar et al. does not first segment the spine before extracting features, but uses a Convolutional Neural Network (CNN) to directly map input images to output fracture classes. They combine a 2D CNN processing sagittal patches along the spine with a Recurrent Neural Network to aggregate predictions of multiple patches from the same patient. While this approach learns features from training data, it only uses 2D sagittal information at a virtually constructed sagittal section to cope with (abnormal) spine curvature [1]. Tomita et al. apply a similar 2D approach using Long Short-Term Memory (LSTM) units for patient-level aggregation [11].

In contrast, in this work we go beyond using learned 2D/2.5D or engineered 3D features by learning compact 3D features for detecting vertebral fractures. The proposed voxel classification method does not require segmenting the spine or (virtually) selecting the appropriate sagittal slice for inspecting vertebral fractures.

2   Data

For this study, we build a training database of 90 de-identified CT image series from the imaging database of the University Hospital of Brussels. These images were acquired on three different scanners (Siemens, Philips and General Electric; 120 kVp tube voltage; maximum in-plane spacing and slice thickness are respectively 0.92 mm × 0.92 mm and 1.5 mm) and contain 90 patients scanned for various indications (average age: 81 years, range: 70–101 years, 64% female patients, 12% negative cases). The dataset has been curated by one radiologist (S.R.) who scored Genant grades (normal, mild, moderate, severe) for every vertebra [5]. It contains a total of 969 vertebrae of which 184 are fractured (85 mild, 64 moderate, 35 severe). Vertebral fracture prevalence is approximately 20% in men and women above 60 years [13] hence our dataset with 19% vertebral fracture prevalence is representative for this clinical population. More than 90% of the scans are abdomen studies implying that more than 75% of the vertebrae range from T11 to S2[1]. Figure 1 shows the number of fractures for every Genant grade along the spine.

---

[1] Vertebrae are named T1 to T12 for thoracic, L1 to L5 for lumbar and S1–S2 for sacral vertebrae (with numbers increasing from top to bottom).

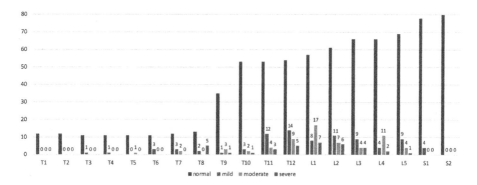

Fig. 1. Training set: number of fractures for every Genant grade along the spine, data labels indicate the amount of mild, moderate and severe fractures.

3 Methods

We present a two-staged vertebra fracture detection method that first predicts a class probability for every voxel using a 3D CNN and secondly aggregates this information to patient-level and vertebra-level fracture predictions. The CT images are resampled to $1\,\mathrm{mm}^3$ and normalized to zero mean and unit standard deviation before voxel classification.

3.1 Voxel Classification

Image classification CNNs map an input image to one output prediction for the entire image. This approach seems attractive for medical imaging as the expert labeling effort is limited to a (set of) answer(s) per image, yet building datasets of (tens of) thousands of CT scans containing subjects of the appropriate classes is not trivial. A voxel classification approach is typically applied for segmentation tasks. While this approach requires much less CT scans, it does require a label for every voxel in each training image which significantly increases the annotation effort. The proposed work applies a hybrid approach combining voxel classification with sparse vertebra annotations. Our experiments demonstrate that this approach produces good results using two orders of magnitude less images than a typical image classification approach.

Since our task is detection and not segmentation, correctly predicting only a *sufficient* amount of voxels around the vertebra centroid is needed to detect normal or fractured vertebrae in an image. We leverage this observation to construct 3D label images for our training database in a semi-automated fashion. First, radiologist S.R. created a text file with annotations for every vertebra present in the field of view as described in Sect. 2. Next, J.N. enriched these labels with 3D centroid coordinates by manually localizing every vertebra centroid in the image using MeVisLab [8]. This step required an average of less than two minutes per image in our dataset. Finally, we extended the method described by Glocker et al. [6] to automatically generate 3D label images from these sparse annotations. The resulting label images contain ellipsoids (flattened along the longitudinal axis for fractured vertebrae) around each vertebra centroid annotated with the ground truth class label provided by the radiologist (combining mild, moderate and severe fractures into one fracture class because of the low number of examples per class, see Fig. 1). The generated label image is not *voxel-perfect* under these assumptions as voxels near the vertebra border are labeled as background in the ground truth label image, but we demonstrate that this is sufficiently accurate for the fracture detection task at hand. The result of this step is a training database $\{(I_k, L_k)\}_{k=1}^{K}$ with K pairs of an image I and label image L of the same spatial dimensions that can be fed into a voxel-based CNN classifier. Note that the above semi-automated procedure is only required for building label images in our training database, test images are processed fully automatically by our method.

3.2   CNN Model Selection

We know that human experts only leverage 2D height information from sagittal slices for detecting vertebral fractures (see Sect. 1), but we want to investigate whether exploiting the 3D information in CT images does yield better results than only using 2D information in the sagittal plane.

Implementation Details. We used the open source Deepmedic Tensorflow implementation by Kamnitsas et al. [7] as this has proven to efficiently sample and process 3D segments from 3D images such as CT (since state-of-the-art GPUs cannot process full 3D image volumes due to memory constraints).

All our experiments have been conducted using the voxel classification network shown in Fig. 2: an 11 layers dual pathway architecture containing 230 K parameters. This CNN consists of 8 convolution layers each of which have filters of size 3^3 realizing an effective receptive field of 17^3 in the normal pathway and 51^3 in the subsampled pathway (subsampling factor 3). This depth has been chosen such that features can be learned using all voxels inside a vertebral body. Additionally, we observed in our experiments that this effective receptive field yields distinct predictions for every vertebra.

The following training regime has been used in all our experiments:

- During training, image segments are sampled in a weighted regime using the ground truth label images to ensure that the network sees enough vertebra voxels. We apply a grid-sampling scheme during inference to build a prediction map of the entire image volume.
- We apply data augmentation by adding noise to our input intensities and randomly flipping images across X, Y and Z axes.
- We use the cross-entropy loss function, RmsProp optimizer, L1 and L2 regularization, anneal our initial learning rate of 0.001 when validation performance plateaus and train for 35 epochs.

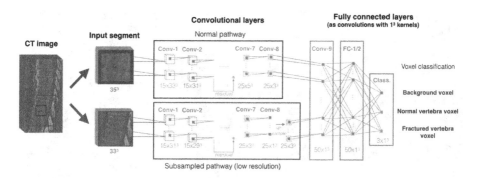

Fig. 2. 3D CNN 11 layers dual pathway architecture realizes an effective receptive field of 17^3 and 51^3 in the normal and subsampled pathway respectively.

Model Selection Experiment. We investigated whether a 3D CNN performs better than a 2D equivalent by comparing three variants of the 11 layer dual pathway network. We split our training database randomly into training (N = 68) and validation (N = 22) to evaluate the three models depicted in Table 1.

Table 1. Model evaluation.

Model name	CONV-1 filter	CONV-2 to 8 filter	Receptive Field (normal pathway)
1slice	$1 \times 3 \times 3$	$1 \times 3 \times 3$	17^2 in sagittal plane
5slices	$5 \times 3 \times 3$	$1 \times 3 \times 3$	17^2 in sagittal plane
3D	$3 \times 3 \times 3$	$3 \times 3 \times 3$	17^3 in 3D

We calibrated all models to contain the same amount of network parameters and trained each model using the training regime described above. Every model has the same effective receptive field in the sagittal plane yet only the 3D model is allowed to learn features outside the sagittal plane. The 5slices model additionally learns to combine information from 5 input slices in the CONV-1 features. Figure 3 qualitatively compares the prediction results of these three models on one validation case. While all three models show similar prediction outputs at coarse scale, the 3D model clearly yields more compact and less noisy predictions than the 2D models at finer scale. All subsequent experiments discussed in this work make use of the 3D model described in this section.

3.3 Aggregation

The voxel classifier transforms an input image into a prediction image that contains a probability $p(f|\mathbf{x})$, $f \in \mathcal{F} = \{\texttt{background}, \texttt{normal}, \texttt{fracture}\}$ for every voxel $\mathbf{x}$ present in the image. This information can be aggregated to patient-level

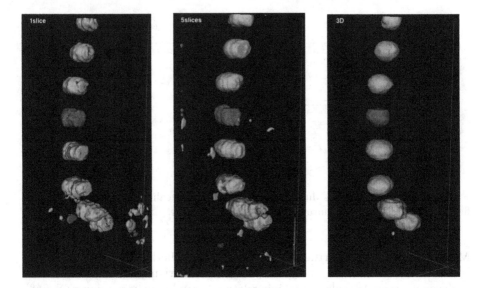

Fig. 3. Prediction outputs for 1slice, 5slices and 3D models on one validation case in our training database. The network explicitly localizes the vertebra and labels each voxel with a class label (green = normal vertebra voxel, red = fractured vertebra voxel). The prediction images shown here are segmentation masks from epoch 30 using hard labels of the most likely class, rendered in 3D with a small counter-clockwise rotation around the longitudinal axis to show prediction results outside the sagittal plane. The 1slice (left) and 5slices (mid) models show stacked (sagittal) predictions. The 1slice (left) and 5slices (mid) models clearly yield more noisy predictions and mixed beliefs inside a vertebra while the 3D (right) model builds more compact predictions. Best viewed in color. (Color figure online)

(detecting whether a fracture is present in the patient image) or vertebra-level (detecting whether a fracture is present for every vertebra visible in the patient image).

Patient-Level Fracture Detection. First, we aggregate the 3D prediction image to patient-level fracture predictions by finding the connected components of fracture voxels and counting the total number of fracture voxels present in the image. This coarse form of aggregation involves two negatively correlated hyperparameters: a *probability threshold* for selecting only those fracture voxels that have been predicted with high probability by our voxel classifier and a *noise threshold* for determining when a component is too small to be a group of vertebra voxels.

Vertebra-Level Fracture Detection. Secondly, we used the ground truth centroid coordinates that were annotated for building our training database (see Sect. 3.1) to perform a more fine-grained aggregation at vertebra-level. The

fracture prediction probabilities of voxels inside a cube around the vertebra centroid are averaged to produce one summary score per vertebra. These probabilities are weighted using a Gaussian distance kernel to decrease the contribution of the voxels further away from the centroid (consistent with our ground truth label images which are *by design* less accurate for voxels that are more distant from the ground truth centroid). For automated screening, we envisage combining our vertebra-level fracture detection method with a vertebra localization method that automatically identifies and localizes each vertebra present in the image. Current state-of-the-art vertebra localization work reports identifying 91.6% of the vertebrae and localizing them with mean error 6.2 ± 16.2 mm [9] on a challenging public dataset. We have used these localization error bounds to add noise to our ground truth centroid coordinates for simulating automated results.

4   Results

We performed a stratified 5-fold cross-validation[2] using 90 images in our training database to estimate the expected performance of our 3D method. For each run, we selected 15% of the images in the training folds as validation samples to determine when to stop training based. We report the Receiver Operating Characteristic (ROC) curve because this metric describes model performance independently of the class distribution and is best suited to compare results from different test sets. The vertebra-level hyperparameter *cube size* has been determined using cross-validation (10 voxels).

Since our **patient-level** fracture detection method involves two hyperparameters that can be chosen to deliver distinct classifiers, we build the ROC curve using the convex hull representing the optimal classifiers from a group of potential classifiers [10]. Each point on this ROC curve represents one optimal classifier generated with one pair of hyperparameter values (probability threshold, noise threshold). Figure 4 shows this patient-level fracture detection ROC curve for the five-fold cross-validation experiment[3]. Our patient-level fracture detection Area Under the Curve (AUC) of 0.95 ± 0.02 is comparable to the results reported by Valentinitsch et al. (AUC 0.88) [12], Tomita et al. (AUC 0.92) [11] and the operating point (recall 0.905, specificity 0.938) on our patient-level fracture detection ROC is similar to the one reported by Bar et al. (recall 0.839, specificity 0.938) [1]. We note that all these results have been reported using different test sets (due to the absence of a public test set for fracture detection). We did not evaluate these other methods on our test set due to the absence of an open source implementation.

---

[2] Since our training database has only 11 negative cases, we stratified the random sampling to ensure that each fold has a minimum of two negative cases.

[3] The (False Positive Rate, True Positive Rate) values have been interpolated to plot a smoother curve.

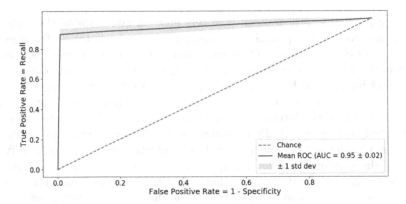

Fig. 4. Patient-level fracture detection ROC curve for the 5-fold cross-validation experiment: bootstrapped (n = 1000) ROC curve (AUC = 0.95 ± 0.02) of our method in blue. (Color figure online)

Figure 5 shows the **vertebra-level** fracture detection ROC curve for the five-fold cross-validation experiment (AUC of 0.93 ± 0.01). We added Gaussian noise to the ground truth vertebra coordinates (standard deviation of 3 mm along each axis) to simulate using automatically detected centroid coordinates (see discussion in Sect. 3.3). We are aware of one vertebral fracture detection work (using a 2.5D method [14]) reporting a sensitivity of 95.7% with a False Positive Rate of 0.29 per patient [4]. Burns et al. designed their test set carefully by excluding cases with more than two contiguous vertebral fractures (in contrast, in our database 18% of the cases contain more than two contiguous fractures and >70% of vertebral fractures have at least one neighboring fractured vertebra, see Fig. 7(b) and (c)). We have not tested this 2.5D method on our test set because of the lack of an open source implementation.

The vertebra-level fracture detection results are illustrated in Fig. 6 (two validation cases with only correct vertebra-level predictions) and Fig. 7 (three validation cases with False Positive (FP) and False Negative (FN) errors at vertebra-level). We observed that our vertebra-level errors occur predominantly on mild cases (either misses on ground truth mild fractures or false alarms on normal vertebrae) and can be clustered into the following categories: errors at edge vertebrae (vertebra and/or its neighbors are not completely visible, see Fig. 7(a)), errors in series of fractured vertebrae (known to be difficult to read as the reference vertebra *disappears*, see Fig. 7(b)) and errors due to confusion with other vertebra pathologies (e.g. inferior vertebra in Fig. 7(b)). Supported by Fig. 1 we hypothesize that our training database does not contain enough vertebra examples, explaining for instance the FN on mild S1 in Fig. 7(a) and the FN on moderate T7 and mild T9 in Fig. 7(c) (notice the spine curvature around L5 and T7–T8 which makes these fractures look different compared to other locations along the spine). We also observed that some ambiguous (mild) cases

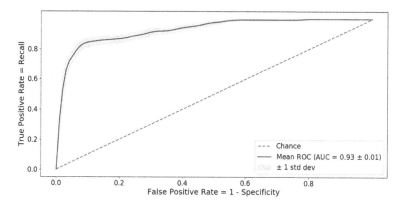

Fig. 5. Vertebra-level fracture detection ROC curve for the 5-fold cross-validation experiment, again bootstrapping (n = 1000) our predictions to generate the mean and standard deviation of ROC curves (AUC = 0.93 ± 0.01).

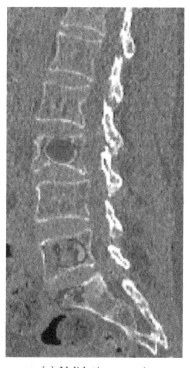

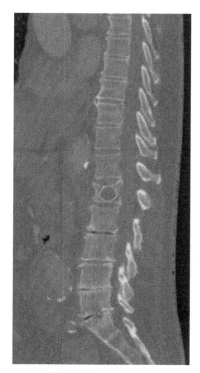

(a) Validation case 1   (b) Validation case 2

Fig. 6. Fracture detection correct vertebra-level predictions on two validation images: for each case, one mid-sagittal slice of the pre-processed 3D input image is overlayed with the output fracture class probability label map (blue = low probability, red = high probability). Probability values <0.05 have been removed for visualization purposes. Best viewed in color. (Color figure online)

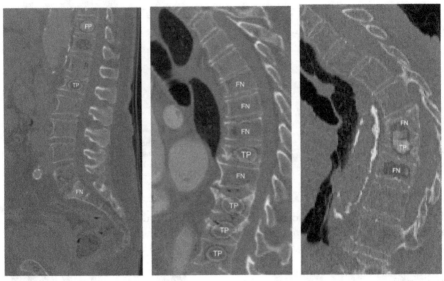

(a) Validation case 3 with two errors at vertebra-level   (b) Validation case 4 with four FNs at vertebra-level   (c) Validation case 5 with two FNs at vertebra-level

Fig. 7. Fracture detection incorrect vertebra-level predictions on three validation images: for each case, one mid-sagittal slice of the pre-processed 3D input image is overlayed with the output fracture class probability label map (blue = low probability, red = high probability), TPs and error type are annotated manually. Probability values <0.05 have been removed for visualization purposes. Observe the differences in image quality and field of view present in our training database. Best viewed in color. (TP = True Positive, FP = False Positive, FN = False Negative) (Color figure online)

would benefit from consensus reading as reported previously [3] (e.g. inferior vertebra in Fig. 7(b)).

5   Conclusion

We present to the best of our knowledge the first vertebral fracture detection model learning 3D features in the spine while simultaneously localizing the detection results to allow for interpretation by radiologists. We discussed the importance of exploiting 3D information to automatically learn compact vertebral fracture detection features. The results of our 5-fold cross-validation experiment demonstrate that our 3D data-driven method produces AUC scores above 90% for patient-level and vertebra-level fracture detection.

While our work demonstrates encouraging fracture detection results, this study has a few limitations which can be mainly attributed to our small database. First, we reported vertebra-level fracture detection results with noisy manual annotations that should be replaced by automatically detected centroids (using an automated localization method). Secondly, we did not yet report fracture grades because our initial experiments show that we have an insufficient number of training examples for many vertebrae and especially for the more ambiguous mild fractures. Thirdly, the amount of thoracic vertebrae and the variability in spine pathologies and image acquisition settings was limited due to the size of our training database. Lastly, we used cross-validation instead of independent training and test sets due to the limited number of patients in our training database.

Acknowledgements. The authors thank the patients, investigators and their teams who took part in this study. The first author is grateful for the comments and feedback provided by Kasper Claes and the discussions on model evaluation with Roberto D'Ambrosio. This study was funded by UCB Pharma and Amgen Inc.

References

1. Bar, A., Wolf, L., Amitai, O.B., Toledano, E., Elnekave, E.: Compression fractures detection on CT. In: SPIE Medical Imaging, pp. 1013440–1013440. International Society for Optics and Photonics (2017)
2. Bromiley, P.A., Kariki, E.P., Adams, J.E., Cootes, T.F.: Fully automatic localisation of vertebrae in CT images using random forest regression voting. In: Yao, J., Vrtovec, T., Zheng, G., Frangi, A., Glocker, B., Li, S. (eds.) CSI 2016. LNCS, vol. 10182, pp. 51–63. Springer, Cham (2016). https://doi.org/10.1007/978-3-319-55050-3_5
3. Buckens, C.F., et al.: Intra and interobserver reliability and agreement of semi-quantitative vertebral fracture assessment on chest computed tomography. PLoS ONE **8**(8), e71204 (2013)
4. Burns, J.E., Yao, J., Summers, R.M.: Vertebral body compression fractures and bone density: automated detection and classification on CT images. Radiology **284**, 788–797 (2017). https://doi.org/10.1148/radiol.2017162100
5. Genant, H.K., Wu, C.Y., van Kuijk, C., Nevitt, M.C.: Vertebral fracture assessment using a semiquantitative technique. J. Bone Miner. Res. **8**(9), 1137–1148 (1993)
6. Glocker, B., Zikic, D., Konukoglu, E., Haynor, D.R., Criminisi, A.: Vertebrae localization in pathological spine CT via dense classification from sparse annotations. In: Mori, K., Sakuma, I., Sato, Y., Barillot, C., Navab, N. (eds.) MICCAI 2013. LNCS, vol. 8150, pp. 262–270. Springer, Heidelberg (2013). https://doi.org/10.1007/978-3-642-40763-5_33
7. Kamnitsas, K., et al.: Efficient multi-scale 3D CNN with fully connected CRF for accurate brain lesion segmentation. Med. Image Anal. **36**, 61–78 (2017)
8. Koenig, M., Spindler, W., Rexilius, J., Jomier, J., Link, F., Peitgen, H.O.: Embedding VTK and ITK into a visual programming and rapid prototyping platform. In: Medical Imaging 2006: Visualization, Image-Guided Procedures, and Display, vol. 6141, p. 61412O. International Society for Optics and Photonics (2006)
9. Mader, A.O., et al.: Detection and localization of landmarks in the lower extremities using an automatically learned conditional random field. In: Cardoso, M.J., et al. (eds.) GRAIL/MFCA/MICGen -2017. LNCS, vol. 10551, pp. 64–75. Springer, Cham (2017). https://doi.org/10.1007/978-3-319-67675-3_7
10. Provost, F.J., Fawcett, T., et al.: Analysis and visualization of classifier performance: comparison under imprecise class and cost distributions. In: KDD, vol. 97, pp. 43–48 (1997)

11. Tomita, N., Cheung, Y.Y., Hassanpour, S.: Deep neural networks for automatic detection of osteoporotic vertebral fractures on CT scans. Comput. Biol. Med. **98**, 8–15 (2018)
12. Valentinitsch, A., et al.: Opportunistic osteoporosis screening in multi-detector CT images via local classification of textures. Osteoporos. Int. **30**, 1275–1285 (2019)
13. Waterloo, S., et al.: Prevalence of vertebral fractures in women and men in the population-based Tromsø study. BMC Musculoskelet. Disord. **13**(1), 3 (2012)
14. Yao, J., Burns, J.E., Wiese, T., Summers, R.M.: Quantitative vertebral compression fracture evaluation using a height compass. In: SPIE Medical Imaging, p. 83151X. International Society for Optics and Photonics (2012)

Metastatic Vertebrae Segmentation for Use in a Clinical Pipeline

Geoff Klein[1,2(✉)], Anne Martel[1,2], Arjun Sahgal[3], Cari Whyne[1,4], and Michael Hardisty[1,4]

[1] Physical Sciences, Sunnybrook Research Institute, Toronto, ON, Canada
[2] Department of Medical Biophysics, Univeristy of Toronto, Toronto, ON, Canada
Geoff.klein@mail.utoronto.ca
[3] Department of Radiation Oncology, University of Toronto, Toronto, ON, Canada
[4] Department of Surgery, University of Toronto, Toronto, ON, Canada

Abstract. Vertebral metastases are common complications of primary cancers that alter bone architecture potentially leading to vertebral fracture and neurological compromise. Quantitative measures from vertebral body segmentations from Computed Tomography (CT) scans have been useful for assessing fracture risk predictions and vertebrae stability. Previous segmentation methods used to generate these metrics were slow and required manual intervention, limiting their utility. More accurate, robust and fast methods are needed for clinical assessments. This investigation proposes a 3D U-Net Convolutional Neural Network (CNN) to accurately segment individual trabecular centrum from metastatically compromised vertebrae of interest in CT imaging. Using different augmentation techniques achieved good performance (DSC = 0.904 ± 0.056) with the segmentation model remaining accurate with simulated lower image quality, and translation of the vertebrae within the image, especially compared to when no augmentations were used (DSC = 0.774 ± 0.188). Integration of this method into a clinical tool will allow accurate and robust quantitative assessment of mechanical stability, aiding clinical decision making to improve patient care.

Keywords: Vertebrae segmentation · 3D U-Net

1 Introduction

Approximately 70% of cancer patients are found to have skeletal metastases during post-mortem examinations [31], with breast, prostate, lung and renal cancers contributing over 80% of cases of skeletal involvement [23]. The spine is the most common site for these skeletal metastases accounting for approximately two-thirds of all skeletal lesions. As cancerous cells metastasize in the spine they can deregulate the balance of bone remodelling, leading to excess bone resorption (osteolytic disease), bone formation (osteoblastic disease) or a mixture of both [23]. Vertebral metastases can cause pain, mobility issues and

© Springer Nature Switzerland AG 2020
Y. Cai et al. (Eds.): CSI 2019, LNCS 11963, pp. 15–28, 2020.
https://doi.org/10.1007/978-3-030-39752-4_2

neurologic compromise. 5–10% of patients will develop mechanical instability or neurological complications requiring treatment. In severe cases further neurlogic complications may cause paralysis.

Local and systemic treatment options exist for patients with spinal metastases. Stereotactic Body Radiation Therapy (SBRT) can deliver more precise highly conformal and hypofractionated external beam radiotherapy with a locally curative objective in single or few fractions. However SBRT has been shown to induce and progress existing vertebral fractures with risks reported ranging from 11% to 39% [4,8,27]. Surgical intervention and stabilization prior to radiation treatment has been considered and shown to improve neurologic outcomes, pain relief, and overall quality of life compared to radiation alone [10]. The spinal instability neoplastic score (SINS) is a scoring system developed to guide clinical assessment of spinal stability and determine if stabilization is necessary [10]. While SINS is in use clinically and shown to be reproducible (0.846 interobserver reproduciblity), it is a semi-quantitative estimation of vertebral stability that is done manually by experts and lacks quantitative metrics of fracture risk or stability.

Quantitative metrics derived from Computed Tomography (CT) scans of vertebral bodies have shown a potent ability to quantify fracture risk. Thibault et al. [30] found that osteolytic tumour involvement $\geq 11.6\%$ relative to the total vertebral body volume was highly correlated to VCF occurring $(P > 0.001)$ with an odds ratio of 37.4 [30] in a cohort of 55 patients. This study was able to predict VCF with a sensitivity and specificity of 70.6% and 94.0%, respectively [30]. Similarly, a pilot study Hardisty et al. found that vertebral volume collapse quantified by analysing CT imaging pre- and post-SBRT was higher in patients that went on to receive mechanical stabilization [12]. These results show promise that a quantitative prediction of spinal stability is possible by evaluating medical imaging. However, these analyses required a multi-step, semi-automated pipeline to segmentation the trabecular centrum and metastatic lesions of a vertebral body [11,33] which was slow, required manual intervention, and was sensitive to initialization (locating and cropping the vertebrae of interest). To evaluate fracture risk and vertebral stability in a clinical setting a faster, automated and more robust segmentation model that is less sensitive to initialization is required. Further the method must be tolerant of variability in scan quality as patients with metastatic disease have imaging studies for treatments planning and follow up with varying resolution. This work proposes that a 3D Convolutional Neural Network (CNN) can be used to meet the necessary criteria for a trabecular centrum segmentation model for clinical use. The pipeline will involve a user selecting a vertebra from a full spine CT scan as the vertebrae of interest will be known. Robustness of the segmentation model will include the ability of it to accurately segment non-centered vertebrae, avoiding the need for careful pre-alignment of the vertebra selected.

2   Dataset

Data from this work was acquired through the Sunnybrook Health Science Centre and consisted of T4-L5 healthy and metastatic vertebrae from 30 patients with sequential imaging taken at four 4-month scan intervals (initial, 4-month, 8-month, 12-month). Fractured vertebrae were excluded from this cohort. Approximately 530 segmentations were used for training and 130 for validation. Ground truth annotations of the vertebral body's trabecular centrum were obtained using the previously developed semi-automated method [11] followed by manual corrections. Each scan was sparsely labelled such that not all vertebra are have ground truth labels.

3   Previous Work

Methods for 3D vertebral body segmentation exists using shape and statistical models, and atlas based methods [5, 11, 13, 14, 16–20, 25, 28]. However, these methods are slow and may only be semi-automated requiring user input. More recent advances have looked into using machine learning techniques for vertebra detection including iterative marginal space learning [24], random forest [6], and Support Vector Machines [9]. These previous segmentation methods require training both a detection and segmentation model, making them less efficient than a single stage model required here. Similarly, the detection model adds additional computational overhead which is unnecessary here as a clinician would already know the vertebrae of interest.

Recent techniques have looked into using classical and convolutional neural networks (CNN) for vertebra segmentation. Al Arif et al. [1] also used a 2D CNN framework to achieve 0.944 Dice similarity coefficient by utilizing shape aware loss function with a U-Net architecture [26]. Janssens et al. [15] was able to accurately label 3D vertebral bodies with a U-Net after vertebra localization with a prior FCN. A single end-to-end model was developed by Lessmann et al. [21] to both segment and classify vertebral bodies in a full CT image using an iterative patch-based method. Sekuboyina et al. [29] demonstrated a whole 3D vertebrae segmentation model that is able to accurately segment vertebrae under varying field-of-view, and with varying pathologies (healthy, osteoporotic unfractured, osteopototic fractured). Even though their model does not achieve higher than state-of-the-art segmentation accuracy (achieving only 0.87 average Dice similarity coefficient) their model can perform under more varying clinical conditions. They achieved this by first cropping 3D vertebrae from CT scans using the localization results of a 2D sagittal attention network. These 3D cropped vertebrae where then segmented using a network composed on both 2D and 3D convolutions. While this work demonstrates a robust segmentation model the 2D convolutions can cause vertebral information to be lost. Also, the two networks are trained and implemented independently and not through an end-to-end structure, meaning final computation and use of the segmentation results in a clinical pipeline will be slower. Furthermore, their approach does seem to significantly decrease in accuracy for osteoporotic vertebrae (Dice coefficient of 0.837) which could cause it's performance on metastatic vertebrae (what is of interest here) to be not clinically usable. A volumetric concurrency of 0.89 has been shown to be sufficient for automatically predicted segmentations to be useful for fracture risk assessment in the metastatically involved spine [11,22,30]. However, other than the semi-automated model by Hardisty et al. [11], none of these models have shown to be able to segment metastatically involved vertebral bodies. It has also not been shown how these models function with the requirements necessary for deployment in the clinical pipeline.

4   Methodology

This work proposes to use a 3D U-Net CNN [7,26] to segment metastatically involved trabecular centrum from CT scans. The method will involve selection of the vertebrae of interest by the user and then automated segmentation and analysis with a 6-layer U-Net (Fig. 1). A U-Net is similar to an encoding (left) and decoding (right) CNN. An image's structural information and resolution decrease as it propagates down the left side of the U-Net as the network attempts to learn high and low level image information, as with normal FCN models. The benefit of the U-Net is the concatenation of the left and right sides which allow the network to regain structural information and resolution is increased during up-convolutions. U-Net models have been shown to be successful for 2D [32] and 3D [21] vertebral segmentation, but have not considered metastatically involved vertebrae. For this work, a 6-layer 3D U-Net was used with a 32 base filters. This was chosen based on the computational resources available and results in existing literature.

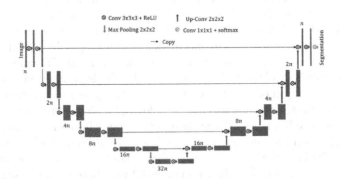

Fig. 1. Example of a 6-layer 3D U-Net CNN with base filter size of n.

4.1   Preprocessing and Training

Prior to training, all CT scans were resampled to $1\,\text{mm}^3$ voxel spacing. Patches (of size $128 \times 128 \times 64$) encompassing individual vertebrae were cropped from the whole-spine CT scans. An example of this patch with and without the ground truth segmentation can be seen in Fig. 2a and b, respectively. Different models were trained based on different augmentation techniques, to investigate their ability to handle simulated real-world situations. Each model was trained using the negative of the Dice Similarity Coefficient (DSC), a measure of overlap for segmentations, for their loss function. The DSC can be seen in (1). Models were developed in Keras and training was done for 200 epochs using the Adam optimizer with a $1e-4$ learning rate. Training was done using 4 NVIDIA Tesla V100 GPUs using Compute Canada resources (https://docs.computecanada.ca/wiki/Beluga/en).

4.2 Augmentation

The goal of this segmentation model is to be used to aid clinical decision making and planning for patients with known disease in specific vertebrae. The user will be able to identify and localize the vertebrae of interest, however the precise location will not be ensured, therefore the segmentation model must segment a vertebral body without precisely knowing the centre. To avoid this necessity translation augmentations were considered during training to shift the cropping window to off-set the vertebral body's center from the windows center. Models trained with this augmentation are denoted as T_M, and an example of this augmentation's affect of the CT data can be seen in Fig. 2c. This segmentation model will also need to be able to segment the vertebral body from CT scans of variable resolution to accommodate different clinical requirements (planning, diagnostic, monitoring) and scanning protocols. A resampling augmentation was used during training, after the initial $1 \, mm^3$ voxel spacing resampling discussed above, to simulate the potential differences in imaging resolution possible in different scans of these patients. Image volumes were resampled with a random voxel spacing between 1–$2 \, mm^3$ and then to the standard $1 \, mm^3$ voxel size used as input to the model. These additional resamplings would decrease image quality in a manner consistent with lower resolution imaging. This range was decided upon based on our groups clinical experience with CT scanning done for planning SBRT compared to those scans done as follow up to the therapy. Models trained with this resampling augmentation are denoted as R_M and an example of this can be seen in Fig. 2d. Finally, shape augmentations common to image processing models were also applied during training and these include randomly rotating, scaling, elastically transforming and mirroring scans. Model's trained with shape augmentations are denoted as S_M.

A total of 8 models were trained based on the combination of the three augmentation techniques (shape, translation and resampling) with the inclusion of a model without any augmentations applied, and were: STR_M, SR_M, ST_M, S_M, TR_M, R_M, and N_M. An example of a of an axial CT slice with both translation and resampling augmentations can be seen in Fig. 2e.

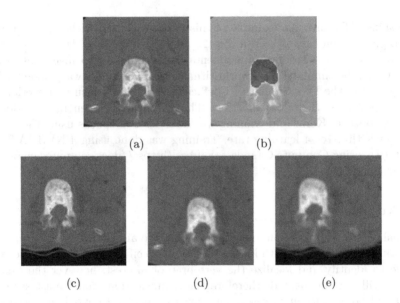

Fig. 2. An axial slice of CT scans showing (a) a centered vertebral body with no augmentations applied; (b) ground truth segmentation overlayed in red; (c) translation augmentation applied; (d) resolution augmentation applied; (e) both translation and resolution augmentations applied. (Color figure online)

4.3 SpineWeb Data

A second analysis was done using the vertebral body segmentations from Dataset 4 from SpineWeb [2,3], using trained model's discussed above. The 2D vertebral body segmentations were concatenated producing 21 3D ground truth samples, of which 7 were excluded as the whole vertebrae was not visible. The resulting 14 samples were also resampled to a consistent $1\,\text{mm}^3$ voxel spacing and patches of $128 \times 128 \times 64$ were cropped as discussed above. These 14 samples were only used for subsequent testing and no training was done with this data.

4.4 Evaluation

Segmentation accuracy was based on the Dice Similarity Coefficient (DSC) (1) and Concurrency (2) which are the harmonic and arithmetic mean of the precision and recall, respectively. The previous semi-automated method achieved an average vertebral body segmentation concurrency of 0.893 ± 0.02 [11]. Therefore, segmentation validation will be based on achieving a minimum average Concurrency of 0.893 for vertebral body segmentation.

$$DSC(X,Y) = 2\frac{|X \cap Y|}{|X| + |Y|} \tag{1}$$

$$Concurrency(X,Y) = \frac{1}{2}\left(\frac{|X \cap Y|}{|X|} + \frac{|X \cap Y|}{|Y|}\right) \tag{2}$$

To determine how each model configuration responded to the different possible real world scenarios being investigated (non-centered vertebral body and low resolution scans) different augmentations were applied to the data during validation. Each model was therefore validated four times for the four different augmentation combinations. Similar to the different model configurations, the four augmentations used during validation were N_D, T_D, R_D, and TR_D. The subscript D is used to identify *validation data* and to distinguish this from a model configuration. The average DSC and Concurrency for the four validation tests will be used to provide an overall evaluation of each model's segmentation accuracy. The DSC provides comparability of these results to other segmentation literature as DSC is one of the most common metrics used for segmentation overlap. Concurrency will provide a comparison to the previous segmentation method by Hardisty et al. (average Concurrency of 0.893%). It is hypothesized that the segmentation model described here is more accurate and more robust than previous semi-automated methods.

5   Results

The DSC and Concurrency results for each model can be seen in Tables 1a and b, respectively, for each validation dataset. Models trained with all augmentations (STR_M) produces the most accurate DSC and Concurrency results of 0.904 and 0.908, respectively, based on the average of the four validation tests. Tables 1a and b show that translation augmentations produce the most significant change to a model's performance, especially when the model is tested on translated data. Shape augmentation produces overall improvement to model performance, but less significant than translation augmentations. Resampling augmentation is shown to only cause, at most, marginal improvements to a model's segmentation accuracy.

The results in Tables 1a and b show that the top-performing model was able to segment the vertebral body more accurately than the previous method by Hardisty et al. (average Concurrency of 0.893), and was able to do this with non-centered vertebrae and under simulated low-resolution conditions. Each individual prediction was computed in approximately 0.4 s using a NVIDIA Titan Xp GPU, which is also approximately 100-times faster than the previous semi-automated segmentation method [11]. A module for deployment of the model has been developed within the 3D Slicer platform that allows users to segment vertebrae of interest [https://github.com/kleingeo/Vertebral_Segmentation].

Examples of axial slices of ground truth and predicted segmentations of the top performing model, STR_M, can be seen in Fig. 3 for the N_D (top) and TR_D (bottom). The DSC and Concurrency values for each example are for the full 3D trabecular centrum. Significant osteoblastic involvement (the T6 example) shows situations where the model has difficulty segmenting the trabecular centrum. This may indicate the need for more examples of this type of tumour involvement or strategies that preserve vertebrae shape [1].

Table 1. DSC, (a), and Concurrency, (b), validation results for the different augmentation and validation tests. Subscripts M and D are used to distinguish augmentations used during training and the validation datasets, respectively. S, T, and R correspond to shape, translation, and resampling augmentations respectively, where N is for no augmentations. *Total Avg.* column is the average DSC or Concurrency for all four test cases.

(a)

Model	DSC (± std)				
	N_D	T_D	R_D	TR_D	Total avg.
STR_M	0.904 (0.048)	0.904 (0.049)	0.904 (0.067)	0.905 (0.060)	**0.904** (0.056)
SR_M	0.910 (0.031)	0.828 (0.215)	0.914 (0.029)	0.820 (0.197)	0.868 (0.154)
ST_M	0.905 (0.038)	0.905 (0.038)	0.897 (0.050)	0.895 (0.044)	0.900 (0.043)
S_M	0.908 (0.033)	0.841 (0.118)	0.904 (0.036)	0.852 (0.092)	0.876 (0.084)
TR_M	0.899 (0.041)	0.899 (0.041)	0.896 (0.042)	0.895 (0.043)	0.897 (0.042)
R_M	0.906 (0.033)	0.754 (0.147)	0.904 (0.033)	0.729 (0.171)	0.823 (0.141)
T_M	0.896 (0.045)	0.896 (0.046)	0.885 (0.083)	0.880 (0.079)	0.889 (0.066)
N_M	0.910 (0.038)	0.688 (0.164)	0.905 (0.039)	0.595 (0.192)	0.774 (0.188)

(b)

Model	Concurrency (± std)				
	N_D	T_D	R_D	TR_D	Total avg.
STR_M	0.908 (0.039)	0.908 (0.039)	0.909 (0.046)	0.909 (0.044)	**0.908** (0.042)
SR_M	0.913 (0.030)	0.844 (0.191)	0.916 (0.027)	0.844 (0.152)	0.879 (0.129)
ST_M	0.908 (0.036)	0.908 (0.036)	0.901 (0.046)	0.899 (0.040)	0.904 (0.040)
S_M	0.912 (0.031)	0.848 (0.115)	0.907 (0.034)	0.859 (0.089)	0.881 (0.081)
TR_M	0.902 (0.038)	0.903 (0.038)	0.900 (0.039)	0.899 (0.040)	0.901 (0.039)
R_M	0.909 (0.031)	0.776 (0.124)	0.908 (0.032)	0.753 (0.150)	0.836 (0.123)
T_M	0.900 (0.041)	0.901 (0.042)	0.893 (0.053)	0.888 (0.053)	0.895 (0.048)
N_M	0.913 (0.035)	0.717 (0.145)	0.908 (0.036)	0.643 (0.169)	0.795 (0.165)

5.1   Results - SpineWeb Data

With the 14 3D vertebral bodies, augmentations were applied in the same manner as with the evaluation above using the trabecular centrum in-house data resulting in a model being tested on four validation datasets. The DSC for the different models on each augmentation test can be seen in Table 2 below. The results here are similar to those above with the in-house data. Translation augmentation shows the most significant change to segmentation accuracy when considering translated data. Resampling augmentation produces minor improvements to overall segmentation accuracy. However, unlike with the in-house data, model's trained with shape augmentations seem to have significantly lower performance relative to the other values in Table 2. Comparing Tables 1a and 2 show

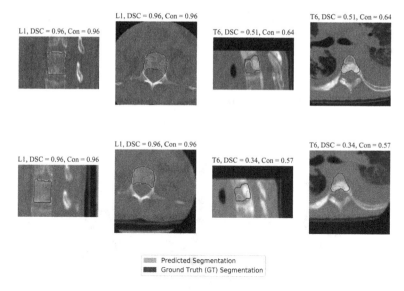

Fig. 3. Best (L1) and worst (T6) trabecular centrum segmentation predictions from STR_M on the N_D (top) and TR_D (bottom). Blue and red contours are the ground truth and predicted segmentations, respectively. (Color figure online)

there is a significant decrease in absolute DSC values and this is easily seen when looking at the N_M and N_D value as this represents the best case scenario (perfectly centered, no changes in resolution or augmentation). An example prediction (red) and ground truth (blue) segmentation can be seen in Fig. 4 below

Table 2. DSC validation results for the different augmentation and validation tests for the SpineWeb Dataset 4 [2,3]. Subscripts M and D distinguish models and the validation datasets, respectively. S, T, and R correspond to shape, translation, and resampling augmentations respectively, where N is for no augmentations. *Total Avg.* column is the average DSC of all four validation datasets.

Model	DSC ($\pm$ std)				
	N_D	T_D	R_D	TR_D	Total avg.
STR_M	0.731 (0.204)	0.728 (0.204)	0.774 (0.030)	0.772 (0.030)	0.751 (0.147)
SR_M	0.468 (0.335)	0.123 (0.214)	0.589 (0.247)	0.077 (0.147)	0.314 (0.329)
ST_M	0.718 (0.201)	0.717 (0.200)	0.695 (0.168)	0.661 (0.209)	0.698 (0.197)
S_M	0.772 (0.036)	0.537 (0.190)	0.760 (0.039)	0.629 (0.209)	0.674 (0.174)
TR_M	0.771 (0.021)	0.769 (0.022)	0.765 (0.021)	0.767 (0.021)	**0.768** (0.021)
R_M	0.780 (0.022)	0.482 (0.203)	0.779 (0.022)	0.468 (0.213)	0.627 (0.212)
T_M	0.774 (0.029)	0.772 (0.027)	0.759 (0.032)	0.756 (0.035)	0.765 (0.032)
N_M	0.784 (0.024)	0.261 (0.0197)	0.771 (0.030)	0.140 (0.159)	0.489 (0.319)

showing both the sagittal and axial views for with the STR_M and N_D. The low DSC and Concurrency reported in Fig. 4 (DSC = 0.77 and Concurrency = 0.81) can be better explained by looking at the predicted and ground truth segmentation contours Fig. 4. The lower DSC was primarily due to the result of the prediction being for the trabecular centrum, rather than the whole vertebral body itself as done in the Spine Web dataset.

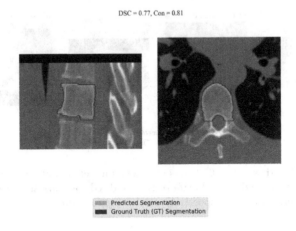

Fig. 4. Sagittal (left) and axial (right) slices of SpineWeb data [2,3] sample without any augmentations with ground truth (blue) and predicted (red) segmentations. Prediction were generated from STR_M, which was trained on trabecular centrum segmentations. The SpineWeb ground truth is of a whole vertebral body segmentation. (Color figure online)

6   Discussion

Tables 1a and b show that one of the highest single DSC value was from N_M on the N_D. This situation constitutes the ideal, real-world scenario as all data is high-resolution and the trabecular centrum is centered in the cropped window. This provides an upper boundary a model's segmentation ability. With a similar argument, the TR_D is the worst case scenario (lower resolution and off-centered cropping) and will show how well the model can handle more difficult situations. The validation results of the N_M with the TR_D achieves the lowest DSC (DSC = 0.595) in Table 1a indicating that the simplest model configuration cannot segment the trabecular centrum accurately for (simulated) real-world situations. The STR_M tested with the TR_D achieves the highest DSC (DSC = 0.905) showing the augmentations have on a model's robustness through its ability to accurately segment the trabecular centrum in real-world situations.

For clinical translation the segmentation model will need to perform well with a variety of image qualities. The R_M can be seen to be able to accurately segment low and high resolution scans, as seen in the N_D and R_D columns in

Tables 1a and b. However, the R_M seems to be only marginally more accurate than the N_M for the same validation datasets. This suggests that the model may be using more shape information in the CT scans to perform the segmentation as opposed to the texture information. Furthermore this may suggest a robustness to variations in image resolution for this segmentation task; the shape of the vertebral bodies was relatively smooth without fine structure. This finding has practical advantages for collecting training data which may not be uniform and for prediction outside of the image resolution of the training set.

Another criteria for clinical use of the segmentation model is for a user to be able to select and segment a vertebral body of interest without needing to have their selection perfectly centered in the cropping window. This will allow for a more robust model that requires less user intervention. Models trained with TA (STR_M, ST_M, TR_M, and T_M) all demonstrate that they can more accurately segment off-centered vertebral bodies compared to models that were not trained with TA. For example the N_M achieved an average DSC of 0.688 ± 0.164 on the T_D, whereas T_M achieved an average DSC of 0.896 ± 0.046. The model trained with TA achieved an approximately 20% greater DSC than the model trained without, as well as a significant reduction in uncertainty. This demonstrates the robustness of the model in its ability to accurately segment off-centered vertebral bodies.

The relative trend of the DSC values with the SpineWeb dataset are similar to the in-house data described above showing that transnational augmentation produces the most significant increase in DSC for translated test cases, improving overall model robustness. The models trained with translation augmentation (T_M), and translation and resampling augmentations (TR_M) performed the highest overall. Unlike the results with the in-house data shape augmentation does not seem to cause any significant improvement and seems to reduce a model's overall DSC, as seen when comparing TR_M and STR_M models. The reason for this can be seen when looking at each sample's segmentation prediction. From the individual results the STR_M model miss-segments one the samples (segments the trabecular centrum of the vertebrae above) resulting in a DSC of 0 and, with only 14 samples, this significantly lowers the average DSC for that scenario. The absolute DSC is decreased for the SpineWeb dataset as seen in Table 2 compared to Table 1a for the in-house data. This can be best seen when comparing the N_M model on the N_D test set, as these represent the best case scenario (perfectly centered, no changes in resolution or augmentation). There are a few potential reasons for this change and they all stem from differences in the physical structures that are segmented in the data. The SpineWeb dataset includes osteophytes in its segmentation which the trabecular centrum data does not. Additionally, the SpineWeb dataset considers the whole vertebral body rather than just the trabecular centrum and this can be clearly seen in Fig. 4. The predicted segmentation in Fig. 4 is qualitatively good if the task was to segment the trabecular centrum, which was what the model was trained for. However, the DSC for Fig. 4 is only 0.77 as it being compared to something slightly different. This does show that the model is able to achieve (qualitiatively) accurate segmentations of the trabecular centrum on different data.

The broad goals of this project are to build tools for the metastatically involved spine to aide in treatment planning, assess mechanical stability, and fracture risk. These methods are of increasing importance with the rising number of treatment options available and the increasing complexity of optimizing care delivery for patients with metastatic disease in the spine. Our future work is focused on tools that can effectively fuse multiple data sources to model the course of symptomatic fractures and indicate the need for prophylactic stabilization. Accurate segmentation of vertebral bodies is the first step for stability estimations. Osteolytic lesion segmentation has previously been done using a patient specific thresholding of Hounsfield units, which required accurate segmentations of the vertebral body [33]. The developed models here (average DSC and Concurrency of 0.904 and 0.908, respectively) were shown to segment the vertebral body accurately enough for this clinical pipeline, and meet the criteria established at the outset of this project based upon this clinical need, Hardisty et al. (average Concurrency of 0.893). The model developed here could be used in a larger clinical pipeline for metastatic fracture prediction and clinical intervention, and could be trained in a complete end-to-end fashion. Personalized cancer care in the spine may also need to consider primary tumour type and local bone quality considerations in building quantitative methods to assess disease progression, vertebral stability and response to therapy.

7   Conclusion

A segmentation model using a 3D U-Net framework has been developed to accurately segment vertebral bodies of metastatically involved vertebrae. Existing methods for vertebral body segmentation in pathologic vertebrae require manual correction and long computational times. This work demonstrates a U-Net framework as a faster, more accurate and robust alternative for segmenting vertebral bodies in 3D CT images of healthy or metastatic vertebrae. Further the network was trained to be robust to changes in scan resolution and positioning of the vertebrae within the scan, making it suitable for different clinical settings and scanning protocols. The proposed segmentation model here can be used in a clinical pipeline and will be able to provide accurate vertebral body segmentation which is a crucial to estimate stability of a patient's spine/vertebrae.

Acknowledgements. Feldberg Chair for Spinal Research, Canadian Institute for Health Research Doctoral Award.

References

1. Al Arif, S.M.M.R., Knapp, K., Slabaugh, G.: Shape-aware deep convolutional neural network for vertebrae segmentation. In: Glocker, B., Yao, J., Vrtovec, T., Frangi, A., Zheng, G. (eds.) MSKI 2017. LNCS, vol. 10734, pp. 12–24. Springer, Cham (2018). https://doi.org/10.1007/978-3-319-74113-0_2

2. Aslan, M.S., et al.: A novel 3D segmentation of vertebral bones from volumetric CT images using graph cuts. In: Bebis, G., et al. (eds.) ISVC 2009. LNCS, vol. 5876, pp. 519–528. Springer, Heidelberg (2009). https://doi.org/10.1007/978-3-642-10520-3_49

3. Aslan, M.S., Shalaby, A., Farag, A.A.: Clinically desired segmentation method for vertebral bodies. In: Proceedings of the International Symposium on Biomedical Imaging, pp. 840–843 (2013)

4. Boehling, N.S., et al.: Vertebral compression fracture risk after stereotactic body radiotherapy for spinal metastases, April 2012

5. Castro-Mateos, I., Pozo, J.M., Pereanez, M., Lekadir, K., Lazary, A., Frangi, A.F.: Statistical interspace models (SIMs): application to robust 3D spine segmentation. IEEE Trans. Med. Imaging 34(8), 1663–1675 (2015)

6. Chu, C., Belavý, D.L., Armbrecht, G., Bansmann, M., Felsenberg, D., Zheng, G.: Fully automatic localization and segmentation of 3D vertebral bodies from CT/MR images via a learning-based method. PLoS One 10(11), e0143327 (2015)

7. Çiçek, Ö., Abdulkadir, A., Lienkamp, S.S., Brox, T., Ronneberger, O.: 3D U-Net: learning dense volumetric segmentation from sparse annotation. In: Ourselin, S., Joskowicz, L., Sabuncu, M.R., Unal, G., Wells, W. (eds.) MICCAI 2016. LNCS, vol. 9901, pp. 424–432. Springer, Cham (2016). https://doi.org/10.1007/978-3-319-46723-8_49

8. Cunha, M.V., et al.: Vertebral compression fracture (VCF) after spine stereotactic body radiation therapy (SBRT): analysis of predictive factors. Int. J. Radiat. Oncol. 84(3), e343–e349 (2012)

9. Dijia Wu, L.L., Lay, N., Liu, D., Nogues, I., Summers, R.M.: Accurate 3D bone segmentation in challenging CT images: bottom-up parsing and contextualized optimization. In: 2016 IEEE Winter Conference on Applications of Computer Vision, WACV 2016, pp. 1–10. IEEE, March 2016

10. Fisher, C.G., et al.: A novel classification system for spinal instability in neoplastic disease. Spine (Phila. Pa. 1976) 35(22), E1221–E1229 (2010)

11. Hardisty, M., Gordon, L., Agarwal, P., Skrinskas, T., Whyne, C.: Quantitative characterization of metastatic disease in the spine. Part I. Semiautomated segmentation using atlas-based deformable registration and the level set method. Med. Phys. 34(8), 3127–3134 (2007)

12. Hardisty, M.R., et al.: Quantitative measures of vertebral body stability in patients developing vertebral compression fractures post-spine stereotactic body radiation therapy: a pilot study. IJROBP (2019, submit)

13. Ibragimov, B., Korez, R., Likar, B., Pernuš, F., Vrtovec, T.: Interpolation-based detection of lumbar vertebrae in CT spine images. In: Yao, J., Glocker, B., Klinder, T., Li, S. (eds.) Recent Advances in Computational Methods and Clinical Applications for Spine Imaging. LNCVB, vol. 20, pp. 73–84. Springer, Cham (2015). https://doi.org/10.1007/978-3-319-14148-0_7

14. Ibragimov, B., Likar, B., Pernus, F., Vrtovec, T.: Shape representation for efficient landmark-based segmentation in 3-D. IEEE Trans. Med. Imaging 33(4), 861–874 (2014)

15. Janssens, R., Zeng, G., Zheng, G.: Fully automatic segmentation of lumbar vertebrae from CT images using cascaded 3D fully convolutional networks. In: Proceedings of the International Symposium Biomedical Imaging, vol. 2018, pp. 893–897. IEEE, April 2018

16. Kadoury, S., Labelle, H., Paragios, N.: Spine segmentation in medical images using manifold embeddings and higher-order MRFs. IEEE Trans. Med. Imaging 32(7), 1227–1238 (2013)

17. Kadoury, S., Labelle, H., Paragios, N.: Automatic inference of articulated spine models in CT images using high-order Markov random fields. Med. Image Anal. **15**(4), 426–437 (2011)

18. Kim, Y., Kim, D.: A fully automatic vertebra segmentation method using 3D deformable fences. Comput. Med. Imaging Graph. **33**(5), 343–352 (2009)

19. Klinder, T., Ostermann, J., Ehm, M., Franz, A., Kneser, R., Lorenz, C.: Automated model-based vertebra detection, identification, and segmentation in CT images. Med. Image Anal. **13**(3), 471–482 (2009)

20. Korez, R., Ibragimov, B., Likar, B., Pernus, F., Vrtovec, T.: A framework for automated spine and vertebrae interpolation-based detection and model-based segmentation. IEEE Trans. Med. Imaging **34**(8), 1649–1662 (2015)

21. Lessmann, N., van Ginneken, B., de Jong, P.A., Išgum, I.: Iterative fully convolutional neural networks for automatic vertebra segmentation and identification. Med. Image Anal. **53**, 142–155 (2019)

22. Leung, J., Hardisty, M., Martel, A., Sahgal, A., Yee, A., Whyne, C.: Convolutional neural networks for vertebral body segmentation in CT images. In: Orthopaedic Research Society, pp. PS1–502 (2018)

23. Maccauro, G., Spinelli, M.S., Mauro, S., Perisano, C., Graci, C., Rosa, M.A.: Physiopathology of spine metastasis. Int. J. Surg. Oncol. **2011**, 1–8 (2011)

24. Michael Kelm, B., et al.: Spine detection in CT and MR using iterated marginal space learning. Med. Image Anal. **17**(8), 1283–1292 (2013)

25. Rasoulian, A., Rohling, R., Abolmaesumi, P.: Lumbar spine segmentation using a statistical multi-vertebrae anatomical shape+pose model. IEEE Trans. Med. Imaging **32**(10), 1890–1900 (2013)

26. Ronneberger, O., Fischer, P., Brox, T.: U-Net: convolutional networks for biomedical image segmentation. In: Navab, N., Hornegger, J., Wells, W.M., Frangi, A.F. (eds.) MICCAI 2015. LNCS, vol. 9351, pp. 234–241. Springer, Cham (2015). https://doi.org/10.1007/978-3-319-24574-4_28

27. Rose, P.S., et al.: Risk of fracture after single fraction image-guided intensity-modulated radiation therapy to spinal metastases. J. Clin. Oncol. **27**(30), 5075–5079 (2009)

28. Ruiz-España, S., et al.: Automatic segmentation of the spine by means of a probabilistic atlas with a special focus on ribs suppression. Med. Phys. **44**(9), 4695–4707 (2017)

29. Sekuboyina, A., Kukačka, J., Kirschke, J.S., Menze, B.H., Valentinitsch, A.: Attention-driven deep learning for pathological spine segmentation. In: Glocker, B., Yao, J., Vrtovec, T., Frangi, A., Zheng, G. (eds.) MSKI 2017. LNCS, vol. 10734, pp. 108–119. Springer, Cham (2018). https://doi.org/10.1007/978-3-319-74113-0_10

30. Thibault, I., et al.: Volume of lytic vertebral body metastatic disease quantified using computed tomography-based image segmentation predicts fracture risk after spine stereotactic body radiation therapy. Int. J. Radiat. Oncol. Biol. Phys. **97**(1), 75–81 (2017)

31. Tseng, C.L., et al.: Spine stereotactic body radiotherapy: indications, outcomes, and points of caution. Glob. spine J. **7**(2), 179–197 (2017)

32. Vania, M., Mureja, D., Lee, D.: Automatic segmentation of spine using convolutional neural networks via redundant generation of class labels. J. Compuat. Des. Eng. Prepr. 1–18 (2017)

33. Whyne, C., et al.: Quantitative characterization of metastatic disease in the spine. Part II. Histogram-based analyses. Med. Phys. **34**(8), 3279–3285 (2007)

Conditioned Variational Auto-encoder for Detecting Osteoporotic Vertebral Fractures

Malek Husseini[1,2](✉), Anjany Sekuboyina[1,2], Amirhossein Bayat[1,2], Bjoern H. Menze[1], Maximilian Loeffler[2], and Jan S. Kirschke[2]

[1] Department of Computer Science, Technical University of Munich, Munich, Germany
malek.husseini@tum.de
[2] Klinikum rechts der Isar, Technical University of Munich, Munich, Germany

Abstract. Detection of osteoporotic vertebral fractures in CT scans is a particularly challenging task that was never sufficiently addressed. This is due to the large variation among healthy vertebrae and the different shapes a fracture could present itself in. In this paper, we combine a *reconstructing* conditioned-variational auto-encoder architecture and a *discriminating* multi-layer-perceptron (MLP) to capture these different shapes. We also introduce a vertebrae-specific loss-weighing regime that maximizes the classification yield. Furthermore, we 'look into' the learnt network by investigating the saliency maps, traversing the latent space and demonstrating its smoothness. Finally, we report our results on two datasets, including the publicly available xVertSeg dataset achieving an F1 score of 84%.

Keywords: Fracture detection · Variational Autoencoders

1 Introduction

The human spine is usually made up from 25 different vertebrae which are divided into 3 sub-categories, i.e. cervical spine vertebrae C1-C7, thoracic spine vertebrae T1-T12 and lumbar vertebrae L1-L5, with the osteoporotic fractures mainly occurring in the thoracic and lumbar regions. The fractures usually present themselves more similarly among similarly-shaped vertebrae. Vertebral fractures are considered one of the main symptoms of osteoporosis and have a multitude of consequences which includes disability and mortality. Furthermore, the mere existence of a fracture exponentially increases the risk of new fractures of the spine and hip and their identification is essential to start the patients bone-protective and bone-enhancing therapy [1]. The prevalence of fractures is high in older adults, reaching 40% by the age of 80 [3]. Nevertheless, under-reporting of incidental fractures remains common with 84% of them not reported in CT exams in one study [4]. The gold standard of fracture classification is currently the semi-quantitative method developed by Genant et. al. [2] which relies on

© Springer Nature Switzerland AG 2020
Y. Cai et al. (Eds.): CSI 2019, LNCS 11963, pp. 29–38, 2020.
https://doi.org/10.1007/978-3-030-39752-4_3

experts judgment to classify and grade the fracture by visually assessing them on a CT scan. These fractures are divided into three levels of severity, as can be seen in Fig. 1, with level 1 being mild to level 3 being severe. We refer to these images as the vertebral masks and are obtained by masking the 3D vertebral patch with its segmentation mask.

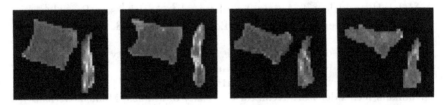

Fig. 1. Levels of severity in vertebral fractures: left to right: the vertebral masks corresponding to the CT mid-slices of a healthy, mildly fractured, moderately fractured and severely fractured L1 vertebrae.

The level of severity is a measure of how much height the vertebral body has lost along the sagittal plane, with mild cases having lost 20–25% of their height, moderate cases 25–40% and the severe case >40%. These scores are meant to be visually interpreted which makes the standard more or less dependent on the observer. Such variation in fractures and their rating complicates the detection task for automated algorithms, keeping in mind that the size of vertebrae vary a lot from one patient to another. Another very common feature among spines with osteoporotic fractures is a decrease in the average vertebral Hounsfield units measured from the CT scan. This is due to the respective decrease of bone density caused by this condition [5]. Thus, to leverage the maximum information out of a scan, an algorithm has to take into account both shape as well as intensity features.

Only a few attempts have been previously made to automate fracture detection and with limited success. A 2D convolutional approach along with LSTMs was used in osteoporotic vertebral fractures to develop a whole-spine CT fracture detection algorithm [6]. However this method requires multiple sagittal slices of the full thoraco-lumbar spine to be visible in the scan, which are rarely available in the clinical setup as most of the scans encompass only a part of the whole spine, this also means that this method is heavily reliant on the quality of those scans and assumes the availability of localized slices of interest. The end result is a binary output of the presence of a fracture in a scan without localizing it to the vertebral level. Thus, making the fracture labels available on a vertebral level would provide the physicians with more precise data to work with.

Moreover, clinically available vertebral data is usually extremely imbalanced with fractures being present in less than 10% of the scans. Therefore, we end up with a large difference between the healthy and fractured classes and training a generalizable supervised model for such cases becomes highly unfeasible. Additionally, the fracture labels are usually only available for a small subset of a huge

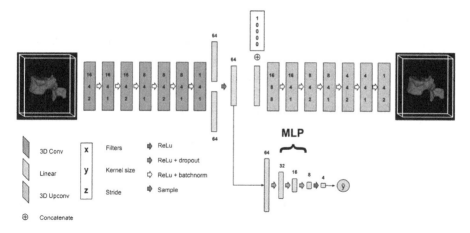

Fig. 2. Architecture of Spine-VAE. The 3-D patches are forwarded through the encoder's convolutional layers, which are separated by batchnorm and ReLU activation layers. The encoder then feeds into two linear layers which represent the mean and the standard deviation through which the latent space is sampled using the gaussian-reparametrization trick. The latent space is concatenated to a vector representing the vertebral classes and then fed to the decoder. If the classification circuit is activated, the latent space vectors are fed to the MLP classifier as well.

un-annotated cohort. To leverage all the data available at hand, unsupervised learning has to be used to capture the underlying differences in the data at hand. To this extent, Autoencoders provide a compressed representation of data based on its successful reconstruction without any supervision. As our problem requires correlating a number of independent features, we look for a frame-work which is able to 'separate' those features as much as possible in the latent representation. Variational Autoencoders (VAEs) provide a framework through which the latent representation can be effectively conditioned. They were used in medical applications to learn interpretable anatomical features for automatic classification of images from patients with cardiac diseases associated with structural remodeling [7].

Our Contributions

1. We propose a compact deep learning architecture which incorporates the unsupervised auto-encoding and the supervised discriminating elements.
2. We use a custom task-specific loss function to learn a representation for our data.
3. We evaluate the model on a publically available dataset.
4. We extensively investigate the interpretability of the acquired results.

2   Methodology

Our approach is to build a two-step deep learning algorithm, that includes the training of an unsupervised model for an un-annotated cohort of vertebrae using a 3D-convolutional VAE. Then, with the help of this model, along with the annotated sub-set of the cohort, we train an MLP to detect vertebral fractures. The network will be referred to as Spine-VAE from now on. In this section we will describe three main parts of our contribution, namely the representation learning, conditioning the VAE, and the MLP. As can be seen in Fig. 2, we use a 3D convolutional VAE to learn the representation of a single vertebra in the latent space. However, we use a couple of tools to enhance our representation and allow it to capture the necessary features to detect a fracture. Our first step is to train the unsupervised model (without the MLP) using all the data-points in our cohort.

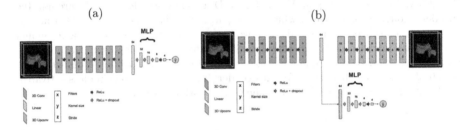

Fig. 3. Baselines architecture. (a) Architecture of the encoder-classifier baseline. (b) Architecture of the normal autoencoder-classifier baseline (AE baseline).

Representation Learning. Since the body region of the vertebra is what should be emphasized on to detect the fractures, we used a weighted mean square error as the loss to train the VAE. As we can see in Eq. 1 the reconstruction loss for a single point of our dataset would be:

$$\mathcal{L}_{recon} = \sum_{i=1}^{n} \omega_i (x_i - \bar{x}_i)^2 \tag{1}$$

where x represents the vertebral mask, $\bar{x}$ is the reconstructed vertebra and the weight ω is a mask of the same size of the vertebral mask and it decreases uniformly from 1 at the beginning of the body to 0 by the end of the vertebral process. An example weight mask can be seen in Fig. 4. As the weights go to zero, the network ignores learning an accurate reconstruction of the down-weighed vertebral process. This results in a very similar reconstruction of the vertebral processes across the different data-points which in turn eases the classification task at the later stage by filtering out a highly variable but fracture-wise unnecessary element of the vertebra. We show the effect of this step on the classification stage in Fig. 5 by plotting the saliency maps from the classifier, back-propagated to the vertebral masks to find which regions were responsible for the classifier's decision.

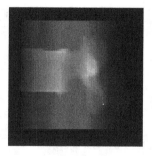

Fig. 4. An example weight mask ω. The red at the very edge of the vertebral body represents 1 which decreases until it reaches 0 (blue) at the end of the vertebral process (Color figure online).

VAE. For learning a latent representation, we use a VAE conditioned on the shape of the vertebrae by concatenating a one-hot vector to the VAE's bottleneck. The one-hot vector is a categorical label obtained by clustering the vertebrae from T1 to L5 into 5 different clusters based on their shape similarities with the groups being as the following: T1-T5, T6-T9, T10-T12, L1-L4 and L5. The idea behind this step is that fractures of vertebrae from the five sub-classes, tend to have more consistent features among their respective classes. Hence, we subtly pass this information to the VAE in order to condition the latent space on that relationship. So we end up with the standard loss function of a VAE:

$$L_{VAE}(\theta, \phi; x_i) = -D_{KL}(q_\phi(z|x_i))||p_\theta(z)) + E_{q_{\phi(z|x_i)}}[log p_\theta(x_i|z)] \qquad (2)$$

where the first term is the KL divergence of the approximate posterior from the prior and the second component is the expected reconstruction error. To balance our losses we include a β term as suggested in [8] and to aid convergence.

MLP. After the VAE converges, we move to the discriminatory part of our network, where the classification circuit is activated by adding a cross entropy loss term to the existing model and our total loss is expressed as:

$$\mathcal{L}_{total} = \mathcal{L}_{recon} + \beta\mathcal{L}_{KL} + \zeta\mathcal{H}(y, \bar{y}) \qquad (3)$$

where ζ is the median frequency weighing map (to handle class imbalance in the dataset) as described in [9], $\mathcal{H}$ is the cross-entropy loss function, y is the ground truth and $\bar{y}$ is the MLP's prediction.

3   Results

In this section, we qualitatively validate our contributions by investigating *saliency maps*, we check the *latent space sanity* by interpolating between different vertebrae and we finally we demonstrate the *descriptiveness* of out latent code. Furthermore we quantitatively analyze our classification results on our in-house as well as xVertSeg datasets.

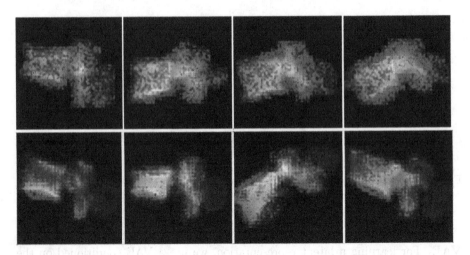

Fig. 5. The contribution of the weighted MSE loss to the classifier demonstrated using the saliency maps of the classified vertebrae projected on the summed vertebral masks along the sagittal axis. Top: saliency maps from a model trained with a standard MSE loss. Bottom: trained with our weighted MSE loss. The heat maps show that the weighted loss decreased the classifier's emphasis on the posterior element of the vertebrae and focused on the body for the decision making process

Data Preparation. Recent advancement in localization and segmentation methodologies made segmentation of the spine and vertebrae easily obtainable from CT scans. In our case, the segmentations were generated using an automated algorithm and corrected by radiologists. Although vertebral segmentations alone provide a valuable insight for fracture detection in terms of structural features, combining those features with their corresponding Hounsfield unit values would provide a classification algorithm with a more complete set of features to correlate with fractures. Therefore, we took the available segmentation masks and multiplied them with the original scans to obtain our 3D masked vertebrae. We re-sampled all of our data to isotropic 1.5 mm voxel spacing and placed our vertebrae in $64 \times 64 \times 64$ boxes.

Datasets and Training. We use a cohort of 2700 segmented vertebrae, among which 1150 are labeled and subsequently 400 fractured samples. Online augmentations are applied where individual samples are rotated randomly to $\pm 30°$ along the sagittal plane. An Adam optimizer was used with a learning rate of 1×10^{-4}. The network was trained for 600 epochs, with the first 150 being trained solely with reconstruction loss while the KL-loss is annealed afterwards with a weight of 0.1 to surmount the problem of over-regularization. The MLP classifier is later trained with cross-entropy loss for 30 extra epochs until convergence. For testing, we used the xVertSeg dataset which has 75 vertebrae. The GPU used was NVIDIA P6000 and training a network required around 10 h.

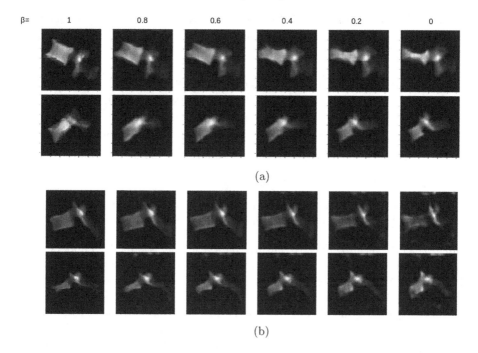

(a)

(b)

Fig. 6. Traversing the latent space. (a) Latent space interpolation: the interpolation between the latent vectors corresponding to two healthy and fractured vertebrae, where the interpolated vector $z = \beta.z_{vert1} + (1 - \beta)z_{vert2}$. Top row: interpolation using Spine-VAE. Bottom row: interpolation using a normal autoencoder. (b) Latent space manipulation: (left to right) top row: 'inducing' a fracture onto a healthy T5 vertebra. Bottom row: 'curing' a level 3 fractured T5 vertebra.

3.1 Qualitative Analysis

Saliency Maps: To investigate the contribution of our custom reconstruction loss in 1, we plot the saliency maps obtained by backpropagation in Fig. 5. We can observe the contrast between the classifiers trained with the standard loss versus the ones trained with our weighted loss, where the decision making process is mainly coming from the body of the vertebrae in the latter, more precisely, from the fractured region. This indicates that this loss constrains the VAE in terms of the features it emphasizes on learning.

Latent Space Sanity: To demonstrate the superiority of VAEs to normal AEs in terms the latent space smoothness, we interpolate between two given vertebrae. In Fig. 6a, we compared the interpolation between Spine-VAE (top) and a normal autoencoder (bottom), with the former producing viable vertebrae reconstructions at all of the interpolated points, while the normal autoencoder resulted in a discontinuous interpolation and distorted images, which cannot exist as a natural variation in reality. In the same experiment, we also observe the interpolation between a healthy and a fractured L1 vertebrae. As can be seen in Fig. 6a

Table 1. Classification statistics. We present the recall, precision, F1 score and the specificity for our three baselines as well as Spine-VAE.

	Model	Recall	Precision	F1	Specificity
In-house dataset	Baseline	62.9 ± 5.3	67.36 ± 2.8	66.4 ± 5.4	67.5 ± 5.5
	AE-baseline	54.3 ± 7.2	52.4 ± 9.2	53.4 ± 8.2	60.3 ± 4.6
	VAE	73.1 ± 3.3	70.3 ± 4.8	71.6 ± 3.2	68.8 ± 6.6
	Spine-VAE	**79.3 ± 2.0**	**71.4 ± 3.1**	**75.4 ± 2.8**	**69.5 ± 4.0**
xVertSeg	Spine-VAE	**85.4**	**83.6**	**84.5**	**66.9**

(top), the transition between the two images is smooth which indicates that the network has learned to encode the different fracture level variations.

Latent Code Descriptiveness. To ensure that our model is capable of understanding the fractures and does the classification process in an interpreted manner, we demonstrate the generation capabilities of our model by *'inducing'* and *'curing'* fractures by manipulating the latent vectors as can be seen in Fig. 6b. Using saliency maps from the classifier to the latent space, we were able to pinpoint the most 'contributing' digits in the latent vector to fracture classification. This is done by determining the two digits with maximum variation between positive and negative samples among the 64 latent space digits. We then varied those digits by adding a bias to them to acquire the induced healthy/fractured states.

3.2   Quantitative Analysis

To validate our approach, we trained 4 networks on our in-house dataset, which are the encoder-classifier baseline as can be seen in Fig. 3a (denoted as 'Baseline'), normal-autoencoder-classifier baseline as seen in Fig. 3b (denoted as 'AE-Baseline'), an ablated version of Spine-VAE which excludes the one-hot vector concatenation and trained with the standard MSE loss (denoted as 'VAE'). Finally, the last network is Spine-VAE. We extracted the recall, precision, F1 score and the specificity for each classifier. All of our results were acquired by 5-fold cross validation on our in-house dataset, with the exception of the xVertSeg results, where we used all the annotated data from our set to train the model and then used the aforementioned set to validate it.

As can been seen in Table 1. We observe an increase in both the recall as well as the overall F1 score as we move from our baseline to the VAEs framework, with the best scores obtained using our Spine-VAE. If we further analyze the results and have a look at the fracture level statistics in Fig. 7a, we observe that the fractures of level 3 are all detected on both our test sets, with level 2 fractures being all detected on the xVertSeg, while having 81% recall on the in-house set. On the other hand, level 1 fractures had less recall with 65% and 80% recalls on the in-house and xVertSeg sets respectively. We can also observe that we have higher recall for level 2 in comparison with level 3 fractures in the 3 trained baseline, an anomaly that was solved in the Spine-VAE classifier.

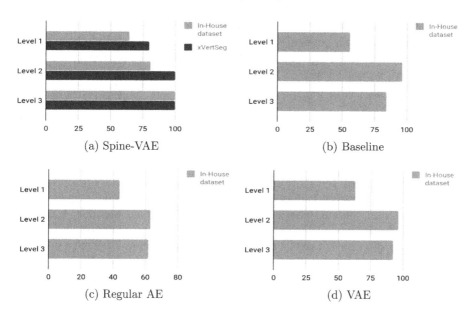

Fig. 7. Recall scores of the different networks for the different fracture levels. A higher recall can be seen for level-2 fractures in comparison with level-3 fractures in the three baselines b, c and d. This anomaly was not present in Spine-VAE.

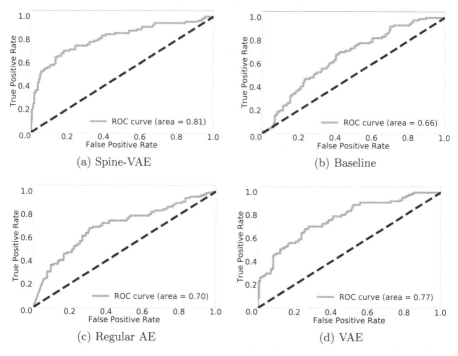

Fig. 8. ROC curves of the different classifiers overall performance along with their respective area under the curve (AUC) values. We can observe a higher AUC for Spine-VAE in comparison with the other baselines.

We also extracted the ROC curves for the different classifiers on our in-house dataset, the results can be seen Fig. 8. We observe an increase in the area under the curve for the 'VAE' and 'Spine-VAE' models with the highest score being 0.81 achieved by Spine-VAE.

4   Conclusions

We proposed a solution to the relatively complicated task of vertebral fracture detection by utilizing the power of deep learning. The presented solution was specifically designed to ensure that the classification process would be based on the relevant structures by ignoring the encoding of the unnecessary posterior elements of the vertebrae as well as establishing a categorical relationship between the fractures and the anatomical type of the vertebrae. Furthermore we managed to overcome the challenge of the limited availability of annotated data by capturing the distribution of the of the un-annotated cohort of vertebrae using VAEs before training a classifier on the available labeled samples. We tested our model on the public xVertSeg dataset to establish the state of the art scores and were able to detect all the clinically relevant cases with 100% recall.

References

1. Williams, A.L., Al-Busaidi, A., Sparrow, P.J., Adams, J.E., Whitehouse, R.W.: Under-reporting of osteoporotic vertebral fractures on computed tomography. Eur. J. Radiol. **69**(1), 179–183 (2009)
2. Genant, H.K., Wu, C.Y., van Kuijk, C., Nevitt, M.C.: Vertebral fracture assessment using a semiquantitative technique. J. Bone Miner. Res. **8**(9), 1137–1148 (1993)
3. Cooper, C., O'neill, T., Silman, A.: European vertebral osteoporosis study group: the epidemiology of vertebral fractures. Bone **14**, 89–97 (1993)
4. Carberry, G.A., Pooler, B.D., Binkley, N., Lauder, T.B., Bruce, R.J., Pickhardt, P.J.: Unreported vertebral body compression fractures at abdominal multidetector CT. Radiology **268**(1), 120–126 (2013)
5. Schreiber, J.J., Anderson, P.A., Rosas, H.G., Buchholz, A.L., Au, A.G.: Hounsfield units for assessing bone mineral density and strength: a tool for osteoporosis management. JBJS **93**(11), 1057–1063 (2011)
6. Tomita, N., Cheung, Y.Y., Hassanpour, S.: Deep neural networks for automatic detection of osteoporotic vertebral fractures on CT scans. Comput. Biol. Med. **98**, 8–15 (2018)
7. Biffi, C., et al.: Learning interpretable anatomical features through deep generative models: application to cardiac remodeling. In: Frangi, A.F., Schnabel, J.A., Davatzikos, C., Alberola-López, C., Fichtinger, G. (eds.) MICCAI 2018. LNCS, vol. 11071, pp. 464–471. Springer, Cham (2018). https://doi.org/10.1007/978-3-030-00934-2_52
8. Higgins, I., et al.: beta-VAE: learning basic visual concepts with a constrained variational framework. In: International Conference on Learning Representations (2017)
9. Roy, A.G., Conjeti, S., Sheet, D., Katouzian, A., Navab, N., Wachinger, C.: Error corrective boosting for learning fully convolutional networks with limited data. In: Descoteaux, M., Maier-Hein, L., Franz, A., Jannin, P., Collins, D.L., Duchesne, S. (eds.) MICCAI 2017. LNCS, vol. 10435, pp. 231–239. Springer, Cham (2017). https://doi.org/10.1007/978-3-319-66179-7_27

Vertebral Labelling in Radiographs: Learning a Coordinate Corrector to Enforce Spinal Shape

Amirhossein Bayat[1,2(✉)], Anjany Sekuboyina[1,2], Felix Hofmann[2], Malek El Husseini[1,2], Jan S. Kirschke[2], and Bjoern H. Menze[1]

[1] Department of Informatics, Technical University of Munich, Munich, Germany
amir.bayat@tum.de
[2] Department of Neuroradiology, Klinikum rechts der Isar, Munich, Germany

Abstract. Localizing and labeling vertebrae in spinal radiographs has important applications in spinal shape analysis in scoliosis and degenerative disorders. However, due to tissue overlaying and size of spinal radiographs, vertebrae localization and labeling are challenging and complicated. To address this, we propose a robust approach for landmark detection in large and noisy images and apply it on spinal radiographs. In this approach, the model has a holistic view of the input image irrespective to its size. Our model predicts the labels and locations of vertebrae in two steps: Firstly, a fully convolutional network (FCN) is used to estimate the vertebrae location and label, by predicting 2D Gaussians. Then, we introduce the Residual Corrector (RC) component, that extracts the coordinates of each vertebral centroid from the 2D Gaussians, and correct the location and label estimations by taking into account the entire image. The functionality of the RC component is differentiable. Thus, it can be merged to the deep neural network, and trained end-to-end with other sub-networks. We achieve identification rates of 85.32% and 52.28% for sagittal and coronal views and localization distance of 4.57 mm and 5.33 mm in sagittal and coronal views radiographs, respectively.

1 Introduction

Localization and labeling in spine is applicable in Cobb angle calculation, surgery planning, bio-mechanical load analysis, diagnosing vertebral fractures or other pathologies. Analysing the shape of the spine in scoliosis and degenerative disorders should be performed in an upright standing position, which is captured using 2D radiographs. Thus, vertebrae localization and labeling is of crucial importance on radiographs, while most of the work for vertebrae labeling has been performed on CT scans.

Vertebrae localization and labeling in radiographs are challenging tasks, due to overlaying tissue and noise. Particularly, in lumbar spine the vertebrae are superimposed by heterogenious soft tissue. Thus the vertebrae's intensity looks

J. S. Kirschke and B. H. Menze—Joint supervising authors.

different from the ones in the other regions. Another challenge is the large size of the spinal radiographs, as they include the entire spine. The latter issue makes it difficult to design a network with a receptive field, covering the entire image. For the same reason, applying available deep neural network architectures to process these images is sub-optimal. On the other hand, applying fully connected layers to address this problem is not feasible as well, as it leads to increasing the parameters, and limits the performance of the network only to fixed size images.

The related works for vertebrae localization and labeling are mostly on CT scans. Usually, an FCN is employed to estimate a heat map of points of interest, then, a second module is applied on the heat map to predict the locations and labels more accurately. The main difference between various methods is mostly in the second module.

In [7] and [8] Yang et al. propose a deep FCN being trained with deep supervision followed by message passing or convolutional LSTM to improve the predictions. In [9] the authors propose to regress a heat map and then regress the landmarks using FCNs. In their approach, the field of view in the input images is variable. While, we always have complete spine image in the input. In [5] and [6] used regression forests label the vertebra, but the limitation of their model is limited field of view. To address the problem of limitation in field of view, the authors in [4] used MLP networks to capture the long-range context features. In the same direction, in [3] the authors introduce an adversarial framework for localization and identification of vertebrae in CT scans. They use an energy-based discriminator in an adversarial setting to correct the labels predicted by the FCN.

While the related works mentioned above, are on CT, we address the problem of vertebrae labeling and localization in spinal radiographs. This task is more challenging due to tissue overlay and noise. We devise a model with unlimited receptive field and enforcing the spinal shape. Our solution is a two-level supervision approach. First, we estimate the rough location of landmarks by predicting 2D Gaussians using a fully convolutional network (FCN). At this level, the 2D Gaussian image is compared to the ground truth 2D Gaussians. Next, we introduce the residual corrector (RC) component, to convert the Gaussians predicted in the previous step, to coordinate format and then correct the coordinates. In this way, the dimensionality of the data decreases from $h \times w$ to 2×24, where h and w are height and width of the input image respectively and 24 is the number of vertebrae. For each vertebra, 2 parameters are estimated as 2D coordinates. In other words, we can decrease the dimensionality of data from input images with variable size to a fixed size of 2×24, as long as the input image includes the entire spine. The functionality of the RC component is differentiable. Therefor, the entire network is trainable in an end-to-end fashion.

Our model design, leverages both the texture features and spinal shape (inter vertebral spatial relation) to localize and label the cervical, thoracic and lumbar vertebrae. We train and test our model on two datasets. One including radiographs of sagittal view and the other one a dataset of coronal view radiograph.

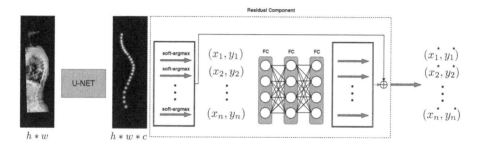

Fig. 1. Overview of the model.

Contributions. (1) We design a robust and accurate deep neural network architecture, with two level of supervision, applicable to landmark detection, on large images. (2) We design residual corrector (RC) component, a differentiable module to convert centroids predicted as 2D Gaussian images to coordinates, and then correct them. (3) Finally, we achieve identification rate of 80.55% for both coronal and sagittal view, localization distance of 7.71 mm and 7.59 mm for sagittal view and coronal view respectively.

2   Methodology

Annotations. The output of the FCN is compared to ground-truth Gaussian images. Similar to [3], the ground-truth Gaussian image $Y \in \mathbb{R}^{(h \times w \times 25)}$ is a 25-channeled, 2D image with each channel corresponding to each of the 24 vertebrae (C1 to L5), and one for the background; h and w are height and width of the input image respectively. Each channel is constructed as a Gaussian heat map of the form $y_i = e^{-||x - \mu_i||^2 / 2\sigma^2}$, $x \in \mathbb{R}^2$ where μ_i is the location of the ith vertebra and σ controls the spread. The background channel is constructed as, $y_0 = -\max(y_i)$.

Architecture. Figure 1 gives an overview of our proposed model. It is composed of two sub-networks: U-net [10] as a FCN, and the RC component, which is highlighted with a dashed-line box in Fig. 1.

In the RC component, first the centroid coordinates are extracted using soft-argmax method [1,2]. In this way, the dimensionality of the data decreases from $h \times w$ to 2×24, where 24 is the number of vertebrae. For each vertebra, 2 parameters are estimated as 2D coordinates. In other words, we can decrease the dimensionality of data from input images with variable size to a fixed size of 2×24, as long as the input image includes the entire spine. After that, we have a residual block of fully-connected layers to (1) correct the location of the estimated coordinates and (2) increase the receptive field of the model. Since the receptive field of the model covers the entire input image, it can capture long-range dependencies in the inputs with fewer network layers and parameters compared to the FCN approaches with similar receptive field. As soft-argmax is differentiable, we can train the fully-connected layers along with the FCN end-to-end and variable size input images. The fully connected layers in RC component sub-network, give a holistic view of the input image to the network and leads to global consistency of the estimated vertebral coordinates and labels. i.e. the order of label and also morphological consistency of the spine.

Training. The FCN sub-network, predicts 2D Gaussians, the predicted image in this level is compared to the ground truth Gaussian image. For this, we use the loss function introduced in [3], which measures the L_2 distance supported by a cross-entropy loss over the softmax excitation of the FCN prediction and ground truth Gaussian image.

$$L_{Gaussian} = ||Y - \tilde{Y}||^2, \tag{1}$$

$$L_{ce} = H(\text{softmax}(Y), softmax(\tilde{Y})), \tag{2}$$

$$L_{img} = L_{Gaussian} + L_{ce}, \tag{3}$$

Where $\tilde{Y}$ is the predicted Gaussian image, Y is the target Gaussian image and H is the cross-entropy function. Next, using our RC component we extract the centroid coordinates and correct them. For this stage, we use a Smooth Absolute (Huber) loss to compare the predicted coordinates to the ground truth.

$$L_{coordinate} = \begin{cases} \frac{1}{2}||C - \tilde{C}||^2 & \text{for } |C - \tilde{C}| \leq 1, \\ \left(|C - \tilde{C}| - 1/2\right) & \text{otherwise.} \end{cases} \tag{4}$$

Finally, the total loss is the sum of the loss at the first and second stages.

$$L_{total} = L_{img} + L_{coordinates}, \tag{5}$$

Inference. Once the model is trained, for inference the input image is fed to the FCN sub-network, the result is a multi-channel image with the same size of the input image. In the resulting image, each vertebral centroid is predicted as a 2D Gaussian in the corresponding channel. Next, this multi-channel image is fed to the RC component. In the RC component, the coordinates of the Gaussians are extracted using the soft-argmax function. Then, the extracted coordinates are corrected by the residual fully-connected layers.

3 Experiments and Results

Datasets. We work with an in-house dataset of 122 patients for each we have the coronal and sagittal radiographs, we also have the vertebral centroids annotations of them. We selected 100 cases randomly for training and 22 cases

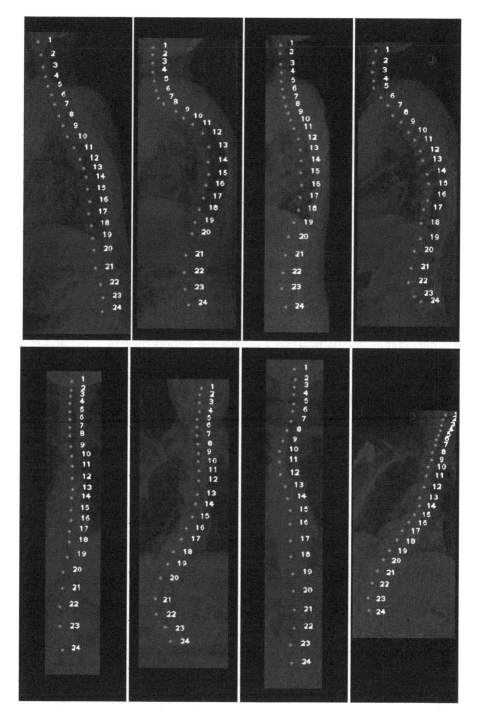

Fig. 2. Testing the model on sagittal coronal radiographs. The predicted centroids are plotted in red dots and the ground truth centroids in green crosses, the numbers indicate the vertebral label. (Color figure online)

for validation. The radiographs are mostly from old patients, some cases with scoliosis or metal insertion. we augmented the data by rotation the images 2, −2, 5 and −5°. We create two separate datasets out of this dataset, one for coronal view and the other one for sagittal view radiographs.

Experiments. We carry out four experiments to validate our approach: (1) First, we train only U-net to localize and label the vertebral centroids on sagittal view. (2) We train U-net on coronal view radiographs. (3) Next, in order to improve the performance of the model, we add the RC component to the model and train the model sagittal view. (4) We train the model with RC component on coronal view radiographs.

We implement our network in the Pytorch framework and use a Quadro P6000 GPU for training the model. In all experiments, the initial learning rate is 0.0001 and Adam is used as optimizer, and the models are trained for 150 iterations.

Evaluation. To evaluate the performance of our network, we use the identification rates (id.rate) and localisation distances (d_{mean} and d_{std}) in mm [5]. Table 1 compares the performance of U-net and our model, trained on sagittal view radiographs. Similarly, Table 2 reports the performance of U-net and our model, trained on coronal view radiographs.

Table 1. Quantitative performance comparison of U-net and our model on sagittal view radiographs.

Measure	U-net	U-net+RC
Id.rate	61.3%	**85.32%**
d_{mean}	11.54 mm	**4.57 mm**
d_{std}	12.73 mm	**3.84 mm**

Table 2. Quantitative performance comparison of U-net and our model on coronal view radiographs

Measure	U-net	U-net+RC
Id.rate	44.5%	**52.63%**
d_{mean}	12.32 mm	**5.33 mm**
d_{std}	11.12 mm	**5.94 mm**

We did not compare our results to vertebrae labeling methods on CT, as they are tested on another modality. The results suggest that the residual correction approach, improves the performance of the model significantly. Due to the limited receptive field of U-net, the localization and labeling are not accurate. The RC component takes the entire spinal shape into account and learn the inter vertebral spatial relation. Therefore, it can correct the predictions of U-net, by enforcing the true shape of spine.

Figure 2 demonstrates examples of testing our model on coronal and sagittal images. The results suggests that our model is robust against metal insertion and treatment effects visible in the radiographs. Also it is robust against shift and small rotations. As it is shown in Fig. 2, our model handles variable size radiographs in both sagittal and coronal view. Finally, our model performs on the raw image data, without preprocessing or employing another network to localize the spine and all of the calculations are done in a single forward pass.

4   Conclusion

Processing large medical images using fully convolutional networks is suboptimal due to limitation of the receptive field and large size of spinal radiographs. Also, it is not feasible to apply fully-connectd layers to these networks to compensate the limitation of receptive fields, as it increases the number of parameters, significantly. Downscaling the images leads to losing details and makes the network prone to error. We propose an architecture to extract the local features using a fully convolutional network and learn the global shape of the spine using a residual corrector module. Finally, our model is robust against treatment effects in radiographs and can localize and label the vertebra in a single forward pass.

References

1. Yi, K.M., Trulls, E., Lepetit, V., Fua, P.: LIFT: learned invariant feature transform. In: Leibe, B., Matas, J., Sebe, N., Welling, M. (eds.) ECCV 2016. LNCS, vol. 9910, pp. 467–483. Springer, Cham (2016). https://doi.org/10.1007/978-3-319-46466-4_28
2. Chapelle, O., Wu, M.: Gradient descent optimization of smoothed information retrieval metrics. Inf. Retrieval **13**, 216–235 (2009)
3. Sekuboyina, A., et al.: Btrfly net: vertebrae labelling with energy-based adversarial learning of local spine prior. In: Frangi, A.F., Schnabel, J.A., Davatzikos, C., Alberola-López, C., Fichtinger, G. (eds.) MICCAI 2018. LNCS, vol. 11073, pp. 649–657. Springer, Cham (2018). https://doi.org/10.1007/978-3-030-00937-3_74
4. Suzani, A., Seitel, A., Liu, Y., Fels, S., Rohling, R.N., Abolmaesumi, P.: Fast automatic vertebrae detection and localization in pathological CT scans - a deep learning approach. In: Navab, N., Hornegger, J., Wells, W.M., Frangi, A.F. (eds.) MICCAI 2015. LNCS, vol. 9351, pp. 678–686. Springer, Cham (2015). https://doi.org/10.1007/978-3-319-24574-4_81
5. Glocker, B., Feulner, J., Criminisi, A., Haynor, D.R., Konukoglu, E.: Automatic localization and identification of vertebrae in arbitrary field-of-view CT scans. In: Ayache, N., Delingette, H., Golland, P., Mori, K. (eds.) MICCAI 2012. LNCS, vol. 7512, pp. 590–598. Springer, Heidelberg (2012). https://doi.org/10.1007/978-3-642-33454-2_73

6. Glocker, B., Zikic, D., Konukoglu, E., Haynor, D.R., Criminisi, A.: Vertebrae localization in pathological spine CT via dense classification from sparse annotations.
   In: Mori, K., Sakuma, I., Sato, Y., Barillot, C., Navab, N. (eds.) MICCAI 2013.
   LNCS, vol. 8150, pp. 262–270. Springer, Heidelberg (2013). https://doi.org/10.
   1007/978-3-642-40763-5_33

7. Yang, D., et al.: Automatic Vertebra labeling in large-scale 3D CT using deep
   image-to-image network with message passing and sparsity regularization. In:
   Niethammer, M., et al. (eds.) IPMI 2017. LNCS, vol. 10265, pp. 633–644. Springer,
   Cham (2017). https://doi.org/10.1007/978-3-319-59050-9_50

8. Yang, D., et al.: Deep image-to-image recurrent network with shape basis learning
   for automatic vertebra labeling in large-scale 3D CT volumes. In: Descoteaux, M.,
   Maier-Hein, L., Franz, A., Jannin, P., Collins, D.L., Duchesne, S. (eds.) MICCAI
   2017. LNCS, vol. 10435, pp. 498–506. Springer, Cham (2017). https://doi.org/10.
   1007/978-3-319-66179-7_57

9. Payer, C., et al.: Integrating spatial configuration into heatmap regression based
   CNNs for landmark localization. Med. Image Anal. **54**, 207–219 (2019)

10. Ronneberger, O., Fischer, P., Brox, T.: U-Net: convolutional networks for biomed-
    ical image segmentation. In: Navab, N., Hornegger, J., Wells, W.M., Frangi, A.F.
    (eds.) MICCAI 2015. LNCS, vol. 9351, pp. 234–241. Springer, Cham (2015).
    https://doi.org/10.1007/978-3-319-24574-4_28

Semi-supervised Semantic Segmentation
of Multiple Lumbosacral Structures on CT

Huaqing Liu[1,2], Haoping Xiao[1], Lishu Luo[3], Chaobo Feng[1,4], Bangde Yin[1,4],
Dongdong Wang[1,4], Yufeng Li[5], Shisheng He[1,4(✉)], and Guoxin Fan[1,6(✉)]

[1] Spinal Pain Research Institute of Tongji University, 301 Yanchang Road, Shanghai, China
862226463@qq.com, {1553058,bondor1553055,
tjhss7418}@tongji.edu.cn, 2496443799@qq.com, 619301798@qq.com,
992273212@qq.com
[2] Artificial Intelligence Innovation Center, Research Institute of Tsinghua, Pearl River Delta,
Guangzhou 510735, China
[3] Tsinghua Shenzhen International Graduate School, University Town of Shenzhen,
Nanshan District, Shenzhen 518055, People's Republic of China
lls18@mails.tsinghua.edu.cn
[4] Shanghai Tenth People's Hospital, Tongji University School of Medicine,
301 Yanchang Road, Shanghai, China
[5] Shanghai Jiao Tong University School of Medicine, 600 Yishang Road, Shanghai 200233,
China
li1554307768@163.com
[6] Department of Spinal Surgery, The Third Affiliated Hospital, Sun Yet-San University,
600 Tianhe Road, Guangzhou, China

Abstract. Labeled data is scarce in clinical practice, and labeling 3D medical data is time-consuming. The study aims to develop a deep learning network with a few labeled data and investigate its segmentation performance of lumbosacral structures on thin-layer computed tomography (CT). In this work, semi-cGAN and fewshot-GAN were developed for automatic segmentation of nerve, bone, and disc, compared with 3D U-Net. For evaluation, dice score and average symmetric surface distance are used to assess the segmentation performance of lumbosacral structures. Another dataset from SpineWeb was also included to test the generalization ability of the two trained networks. Research results show that the segmentation performance of semi-cGAN and fewshot-GAN is slightly superior to 3D U-Net for automatic segmenting lumbosacral structures on thin-layer CT with fewer labeled data.

Keywords: Semi-supervised segmentation · Lumbosacral structures · Semi-cGAN · Fewshot-GAN

1 Introduction

Three-dimensional (3D) radiographic image is beneficial for preoperative assessment of spinal surgery [1]. Computed tomography (CT) excels at demonstrating bony structures,

---

H. Liu and H. Xiao—These two authors equally contribute to the study.

© Springer Nature Switzerland AG 2020
Y. Cai et al. (Eds.): CSI 2019, LNCS 11963, pp. 47–59, 2020.
https://doi.org/10.1007/978-3-030-39752-4_5

while magnetic resonance imaging (MRI) is good at differentiating soft tissues. However, the combined evaluation of lumbosacral structures on multimodal radiographic images for surgical planning is time-consuming and costly. Thin-layer CT is an ideal candidate for constructing a 3D model with bony structures and soft tissue for surgical planning. However, manual segmentation is cumbersome for 3D reconstruction.

Deep learning is a promising technology to achieve the automatic segmentation of medical images. However, labeled data is scarce in clinical practice, and labeling 3D medical data is time-consuming. In 2015, a concise neural network named U-Net was developed for the semantic segmentation of two-dimensional biomedical images with a few training data [2]. One year later, the same group developed 3D U-Net for volumetric segmentation [3]. U-Net or 3D U-Net has been validated as a successful supervised neural network for segmenting medical images [4–8]. Generative adversarial networks (GAN) can integrate unlabeled data to achieve precise segmentation [9].

The study aims to develop a conditioned GAN for 3D images (semi-cGAN and fewshot-GAN) with a few labeled data and investigate its segmentation performance of lumbosacral structures on thin-layer CT.

2   Methods

The local institutional ethical committee approved this retrospective study before data extraction. All the CT data was obtained from Shanghai Tenth People's Hospital, and algorithms were mainly developed and tested using Tensorflow on a personal computer (GPU: An Nvidia Tesla M40 GPU with 11180 MB memory CPU: An Intel(R) Xeon(R) CPU E5-2682 v4 @ 2.50 GHz CPU with eight cores).

2.1   Manual Annotations

Thirty-one cases of thin-layer CT were manually segmented with Slicer 4.8. Lumbosacral nerves, bones and nerves were meticulously segmented and labeled (see Fig. 1). These manual annotations were regarded as the ground truth.

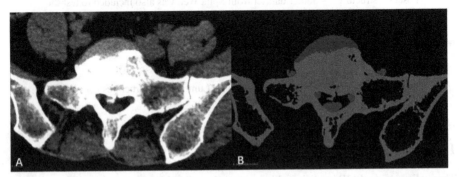

Fig. 1. Manual segmentation of lumbosacral structures. A: original CT with segmentation; B: segmented masks.

2.2 Data Preprocessing

All thin-layer CT were preprocessed using the following steps: resampling, cropping, and intensity normalization.

1. Resampling: we standardized the voxel size of the original CT and manually labeled 3D masked images into 1 mm × 0.5 mm × 0.5 mm nearest-neighbor interpolation.
2. Cropping: we cropped the resampled CT and manually-labeled 3D masked images into 32 × 64 × 64 patches, which were used as input data.
3. Intensity normalization: For more accurate semantic segmentation, the average brightness and contrast fluctuations of CT images of different samples should have a degree of consistency. For this reason, the CT images in the dataset are standardized so that the pixel values of the CT images have zero mean and unit variance. The scale and bias were obtained by statistic computing from the training dataset, and they were used for whitening during all phases, including training, validation and testing.
4. Data augmentation: we conducted data augmentation with the following methods:

   - Adding a small amount of white noise to patch which will be input to the neural network
   - Performing vertical flipping and horizontal flipping at a certain probability
   - Voxel size is randomly disturbed within the range of ±0.2 mm to introduce a degree of size variation
   - Performing random rotation around the Height axis with angles within ±10°.

2.3 Network Architecture

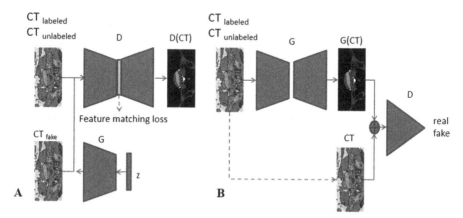

Fig. 2. Schematic drawing of two network architecture. A: Fewshot-GAN network architecture; B: Semi-cGAN network architecture

As Fig. 2 shown, generator G of fewshot-GAN is a volume generator proposed by Wu et al. [10] The input for generator G is a noise vector z and the output of generator

G is a fake patch. In association with a labeled patch and an unlabeled patch, they are input into Discriminator D. Discriminator D is a modified 3D U-Net with leaky ReLUs, weight-normalization (rather than batch normalization) and average pooling (instead of Max pooling). Discriminator D serves as a segmentation network to gain corresponding masks. Features of generated images match features in an intermediate layer of discriminator by feature matching loss.

Generator G of semi-cGAN is a 3D U-Net. The input for generator G is CT patch x (labeled CT or unlabeled CT), and the output of generator G is the corresponding masks of automatic segmentation. G(x) has four channels (4 classes: nerve, bone, disc, background), which is the same to manually labeled masks y (unlabeled CT has no corresponding masks y). G(x) or y concatenate to x, which generates (x, G(x)) or (x, y) and is input to Discriminator D. Discriminator D is a 3D convolutional neural network (3D-CNN). The last two layers are full connection layer, and the last layer generates the probability of (x, G(x)) or (x, y) as a real sample.

2.4 Material

Thirty-one labeled cases and 54 unlabeled cases of thin-layer CT were included for semi-supervised semantic segmentation. The ratio of training labeled cases and validation labeled cases for both semi-cGAN and fewshot-GAN is 20:6. The remaining 5 cases serve as a testing function. As a comparison, the same data used for training 3D U-Net except unlabeled cases. Besides, another ten labeled cases from SpineWeb [11] were also included to test the generalization ability of the two trained networks.

2.5 Training

Fewshot-GAN. This method consists of a generator and a discriminator.

Generator Loss. We adopted the Feature Matching (FM) loss for the generator. This method is aimed at matching the expected value of features $f(x)$ in an intermediate layer of the discriminator.

$$L_{G,labeled} = \left\| E_{x,y:P_{labeled}(x)} f(x) - E_{z:noise} f(G(z)) \right\|_2^2 \tag{1}$$

Discriminator Loss. The discriminator loss function is the sum of three losses incorporating a weighted loss for labeled images.

$$L_D = L_{D,fake} + L_{D,unlabeled} + \lambda_D L_{D,labeled} \tag{2}$$

The loss for labeled images is weighted cross-entropy loss. Different classes have unique values. In terms of unlabeled data, we constraint the output to correspond to one of the K classes of real data. Similarly, to calculate the loss for fake images, we impose each pixel of an input image to be predicted as fake.

$$L_{D,fake} = -E_{z:noise} \sum_{x \in \Omega} log \left[\frac{1}{F(G(z)) + 1} \right] \tag{3}$$

$$L_{D,unlabeled} = -E_{x:P_{unlabeled}(x)} \sum_{x\in\Omega} \log\left[\frac{F(x)}{F(x)+1}\right] \tag{4}$$

$$L_{D,labeled} = -E_{x,y:P_{labeled}(x,y)} \sum_{x\in\Omega} W(X) \log\left[\frac{exp\,(a_1(x))}{\sum_{k=0}^{N} exp(a_k(x))}\right] \tag{5}$$

where $F(x) = \sum_{k=0}^{N} \exp(a_k(x))$, $a_k(x)$ is the activation value at the last layer of the discriminator of voxel point x in channel k; $a_l(x)$ is the activation value at the last layer of voxel point x in the ground truth channel; $w(x)$ is the cross entropy weight of voxel x.

Semi-cGAN. This method also includes a generator and a discriminator.

Generator Loss. The generator is aimed to fool the discriminator and to be near the ground truth in a weighted softmax cross-entropy sense. We made two main modifications to the standard Generator Loss. The first one is including unlabeled data to improve the generalization, and the other is adding a weighted softmax cross-entropy loss for labeled data. Therefore, the final formula is as follow:

$$L_G = L_{G,fake} + L_{G,unlabeled} + \lambda_G L_{G,labeled} \tag{6}$$

Among which:

$$L_{G,fake} = -E_{x:P_{labeled}(x)}\left[\log D(x, G(x))\right] \tag{7}$$

$$L_{G,unlabeled} = E_{x:P_{unlabeled}(x)}\left[\log D(x, G(x))\right] \tag{8}$$

$$L_{G,labeled} = -E_{x,y:P_{labeled}(x,y)} \sum_{x\in\Omega} w(x)\log\left[\frac{exp(a_l(x))}{\sum_{k=0}^{N} exp(a_k(x))}\right] \tag{9}$$

In the above function, $x \in \Omega$ is voxel point of labeled CT; $a_k(x)$ is the activation value at the last layer before softmax of voxel point x in channel k; $a_l(x)$ is the activation value at the last layer before softmax of voxel point x in the ground truth channel; $w(x)$ is the cross-entropy weight of voxel x.

Discriminator Loss. The loss function of the discriminator is as follows:

$$L_D = \lambda_D L_{D,fake} + L_{D,unlabeled} + L_{D,labeled} \tag{10}$$

Among which:

$$L_{D,fake} = -E_{x:P_{labeled}(x)}\left[\log(1 - D(x, G(x)))\right] \tag{11}$$

$$L_{D,unlabeled} = -E_{x:P_{unlabeled}(x)}\left[\log(1 - D(x, G(x)))\right] \tag{12}$$

$$L_{D,labeled} = -E_{x,y:P_{labeled}(x,y)}\left[\log D(x, y)\right] \tag{13}$$

Training Process. We adopted the Adam algorithm to optimize generator loss and discriminator loss. During training, the batch size of labeled data and unlabeled data is 4. It took 48.01 h to complete the training with 100 thousand iterations for semi-cGAN and 70.52 h for fewshot-GAN with the same number of iterations. To compared with state-of-art algorithms, we also trained a 3D U-Net with 20 labeled cases. During the training process, we preserve the best model according to the highest dice score on the testing dataset.

2.6 Testing

A total of 5 labeled cases were used to test the trained semi-cGAN, fewshot-GAN and 3D U-Net. Another 10 cases from SpineWeb were also included to test the generalization ability of the two trained networks. During testing, any case in the test dataset will be selected and undergo standard processing, and then a sliding window of size $32 \times 64 \times 64$ is used to traverse the case with stride $= (20 \times 40 \times 40)$ to obtain the patch x_i. The patch x_i will be input to the trained model M, and then the model will generate the corresponding probability mask y_i. Finally, the automatically segmented mask S will be obtained with the above-mentioned combined algorithm combining y_i based on the location of x_i on Y.

3   Results

3.1   Own Dataset

Final segmentation performances on five cases are compared to each other (see Fig. 3) Confusion matrixes are also calculated to display a summary of prediction results (see Fig. 4). The Dice score of semi-cGAN and fewshot-GAN is 91.5117% and 90.0431%, respectively, compared with 89.8137% of 3D U-Net (see Table 1). The average ASD of semi-cGAN and fewshot-GAN is 1.2726 and 1.5188, compared with 1.4747 of 3D U-Net (see Table 2).

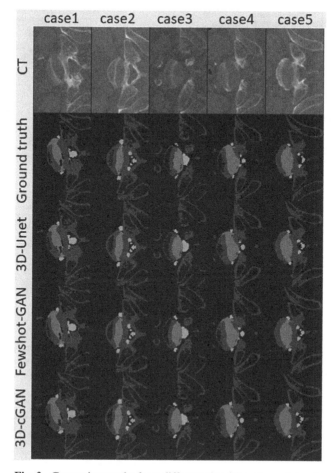

Fig. 3. Generating masks from different algorithms (own dataset).

3.2 Cross Dataset

Similarly, ten cases from the cross dataset are applied to assess segmentation performances (see Fig. 5), and confusion matrixes will be shown in Fig. 6. The Dice score of semi-cGAN and fewshot-GAN is 89.0644% and 88.3881% respectively, compared with 88.6382% of 3D U-Net (see Table 3). The average ASD of semi-cGAN and fewshot-GAN is 0.6869 and 0.6954, compared with 1.3109 of 3D U-Net (see Table 4).

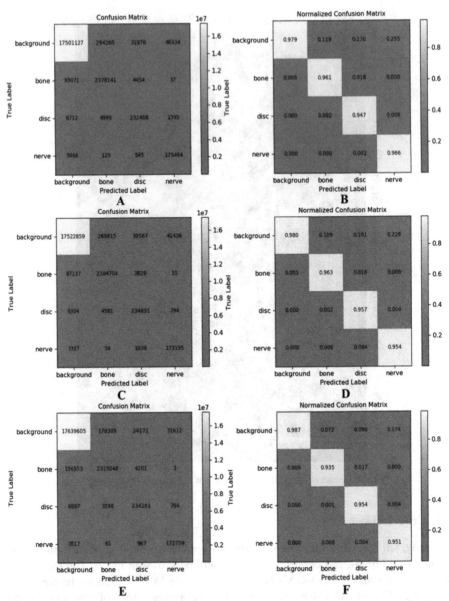

Fig. 4. Matrix of pixel classification (own dataset). A, C, E are confusion matrixes for 3D U-Net, Fewshot-GAN, Semi-cGAN respectively. B, D, F are normalized confusion matrixes for 3D U-Net, Fewshot-GAN, Semi-cGAN respectively

Table 1. Dice score of testing cases from local dataset.

Dice score	Bone	Disc	Nerve	Average
3D U-Net	92.3489	90.5965	86.4957	89.8137
Fewshot-GAN	92.9376	90.1145	87.0771	90.0431
Semi-cGAN	93.2424	92.1263	89.1665	91.5117

Table 2. Average symmetric surface distance of testing cases from local dataset.

ASD\voxel	Bone	Disc	Nerve	Average
3D U-Net	0.3865	3.2963	0.7412	1.4747
Fewshot-GAN	0.3578	3.4492	0.7493	1.5188
Semi-cGAN	0.3275	2.9085	0.5817	1.2726

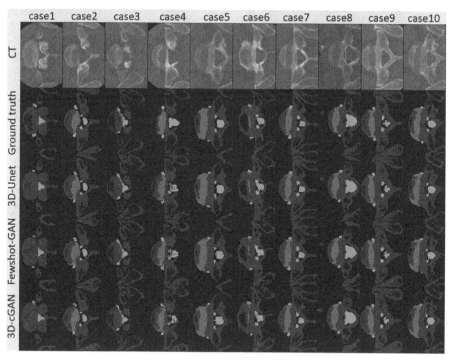

Fig. 5. Generating masks from different algorithms (cross dataset).

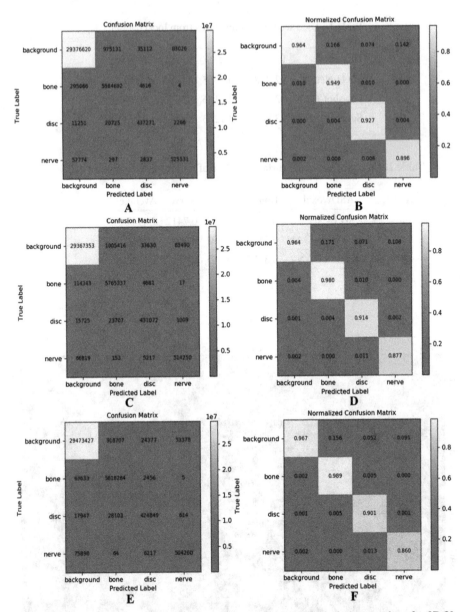

Fig. 6. Matrix of pixel classification (cross dataset). A, C, E are confusion matrixes for 3D U-Net, Fewshot-GAN, Semi-cGAN respectively. B, D, F are normalized confusion matrixes for 3D U-Net, Fewshot-GAN, Semi-cGAN respectively

Table 3. Dice score of testing cases from cross dataset.

Dice score	Bone	Disc	Nerve	Average
3D U-Net	88.3997	90.3211	87.1937	88.6382
Fewshot-GAN	89.5538	87.8308	87.7797	88.3881
Semi-cGAN	90.5567	88.9054	87.7312	89.0644

Table 4. Average symmetric surface distance of testing cases from cross dataset.

ASD\voxel	Bone	Disc	Nerve	Average
3D U-Net	0.6948	0.5044	2.7335	1.3109
Fewshot-GAN	0.6324	0.6524	0.8013	0.6954
Semi-cGAN	0.5507	0.5750	0.9350	0.6869

4 Discussion

Based on enormous raw data, models trained by deep learning network are able to extract pertinent features and accomplish complex tasks. Therefore, relevant clinical data is essential for this promising method to train a prospective model. [12] However, specific task entails corresponding data, and public datasets are usually hard to meet particular research requirements. Besides, labeled data is scarce in clinical practice, and labeling 3D medical data is time-consuming. Apart from these problems, manual segmentation is not always consistent or entirely precise due to inter-expert variance. [13] It is also time-consuming to examine the accuracy of collected data. All in all, accurate and satisfying clinical data is rare, which is a bottleneck of this technology.

Transfer learning is one of the possible methods to remedy the scarce data problem. It applies knowledge learned from other similar data into target domain, which means little target data is required. However, this technology is not a simple task to implement, as the potential similarities between source data distribution and target data distribution need to be identified. This issue may result in negative transfer, which means reduced performance. [14] Because of little publicly related clinical data, this method can not perfectly tackle our problems.

Unsupervised learning is another sensible approach, making its inferences based on unlabeled data. However, unsupervised learning demands more training unlabeled data than supervised learning. It is also time-consuming to obtain relevant data and clean them. Meanwhile, unsupervised learning may show a poor performance in segmentation with lower accuracy because of unlabeled data. Also, given the subjectivity of unsupervised learning and its lack of gold standard, this method can not fix the existing challenges.

Semi-supervised learning subtly utilizes a combination of labeled data and unlabeled data. In this case, few labeled data and relatively more unlabeled data are required. One semi-supervised method proposed by Wenjia Bai et al. [15] has applied to MR Image, which proved the potentiality of semi-supervised learning. Other successful applications like Li et al. [16] and Papandreou et al. [17] showed a promising prospect. Therefore, we adopted semi-supervised learning to tackle the scarcity of clinical data.

U-NET [2] is a well-known convolutional neural network applied to medical image segmentation. This network creatively combined contraction path and symmetric expanding path, which eventually facilitated both segmentation speed and performance. Besides, it has been widely applied to various specific issues and proved its excellent segmentation ability. GANs are a semi-supervised method which has obtained great achievement in medical image processing despite lack of adequate labeled data. By synthesizing realistic images to alleviate the data limitation, this efficient network has an outstanding performance in segmentation. Other than this merit, GANs are highly effective in extracting meaningful features from training data compared with conventional approaches [18].

Both our methods try to combine the benefits of U-NET and GANs. In the fewshot-GAN, the discriminator is a modified 3D U-NET, and the generator is a volume generator. Semi-cGAN consists of a generator based on 3D U-NET and a 3D CNN discriminator. 3D U-NET acts as a different role in two models, and one of our goals is trying to explore which combination has better performance. Recently, there is also a novel U-NET-GAN proposed by Dong et al. However, in their proposed model, three-label-based segmentations are included and trained each of them separately, which might need massive computational powers. Our trained models are more concise with promising results, which automatically segment multiple lumbosacral structures.

5   Conclusions

Segmentation performance of semi-cGAN or fewshot-GAN is slightly superior to 3D U-Net for automatic segmenting lumbosacral structures on thin-layer CT with fewer labeled data.

Acknowledgement. We would like to thanks William M. Wells in Brigham and Women's Hospital, Harvard Medical School for the guidance of this project.

References

1. Kochanski, R.B., Lombardi, J.M., Laratta, J.L., Lehman, R.A., O'Toole, J.E.: Image-guided navigation and robotics in spine surgery. Neurosurgery **84**, 1179–1189 (2019)
2. Ronneberger, O., Fischer, P., Brox, T.: U-Net: convolutional networks for biomedical image segmentation. In: Navab, N., Hornegger, J., Wells, William M., Frangi, Alejandro F. (eds.) MICCAI 2015. LNCS, vol. 9351, pp. 234–241. Springer, Cham (2015). https://doi.org/10.1007/978-3-319-24574-4_28
3. Çiçek, Ö., Abdulkadir, A., Lienkamp, S.S., Brox, T., Ronneberger, O.: 3D U-Net: learning dense volumetric segmentation from sparse annotation. In: Ourselin, S., Joskowicz, L., Sabuncu, M.R., Unal, G., Wells, W. (eds.) MICCAI 2016. LNCS, vol. 9901, pp. 424–432. Springer, Cham (2016). https://doi.org/10.1007/978-3-319-46723-8_49
4. Wang, C., Macgillivray, T., Macnaught, G., Yang, G., Newby, D.: A two-stage 3D Unet framework for multi-class segmentation on full resolution image (2018)

5. Funke, J., et al.: Large scale image segmentation with structured loss based deep learning for connectome reconstruction. IEEE Trans. Pattern Anal. Mach. Intell. **41**, 1669–1680 (2018)
6. Norman, B., Pedoia, V., Majumdar, S.: Use of 2D U-Net convolutional neural networks for automated cartilage and meniscus segmentation of knee MR imaging data to determine relaxometry and morphometry. Radiology **288**, 177–185 (2018)
7. Weston, A.D., et al.: Automated abdominal segmentation of CT scans for body composition analysis using deep learning. Radiology **290**, 669–679 (2019)
8. Huang, Q., Sun, J., Ding, H., Wang, X., Wang, G.: Robust liver vessel extraction using 3D U-Net with variant dice loss function. Comput. Biol. Med. **101**, 153–162 (2018)
9. Dong, X., et al.: Automatic multi-organ segmentation in thorax CT images using U-Net-GAN. Med. Phys. **46**, 2157–2168 (2019)
10. Wu, J.J., Zhang, C.K., Xue, T.F., Freeman, W.T., Tenenbaum, J.B.: Learning a probabilistic latent space of object shapes via 3D generative-adversarial modeling. In: Advances in Neural Information Processing Systems, vol. 29 (2016)
11. Ibragimov, B., Likar, B., Pernuš, F., Vrtovec, T.: Shape representation for efficient landmark-based segmentation in 3-D. IEEE Trans. Med. Imaging **33**, 861–874 (2014)
12. Chen, D., et al.: Deep learning and alternative learning strategies for retrospective real-world clinical data. Npj Digit. Med. **2**, 43 (2019)
13. Retter, F., Plant, C., Burgeth, B., Botella, G., Schlossbauer, T., Meyer-Bäse, A.: Computer-aided diagnosis for diagnostically challenging breast lesions in DCE-MRI based on image registration and integration of morphologic and dynamic characteristics. EURASIP J. Adv. Sig. Process. **2013**, 157 (2013)
14. Pan, S.J., Yang, Q.A.: A survey on transfer learning. IEEE Trans. Knowl. Data Eng. **22**, 1345–1359 (2010)
15. Bai, W., et al.: Semi-supervised learning for network-based cardiac MR image segmentation. In: Descoteaux, M., Maier-Hein, L., Franz, A., Jannin, P., Collins, D.L., Duchesne, S. (eds.) MICCAI 2017. LNCS, vol. 10434, pp. 253–260. Springer, Cham (2017). https://doi.org/10.1007/978-3-319-66185-8_29
16. Li, X., Yu, L., Chen, H., Fu, C.-W., Heng, P.-A.: Transformation consistent self-ensembling model for semi-supervised medical image segmentation (2019)
17. Papandreou, G., Chen, L.C., Murphy, K.P., Yuille, A.L.: Weakly- and semi-supervised learning of a deep convolutional network for semantic image segmentation. In: IEEE International Conference on Computer Vision, pp. 1742–1750 (2015)
18. Kazeminia, S., et al.: GANs for medical image analysis (2018)

AASCE Challenge

Accurate Automated Keypoint Detections for Spinal Curvature Estimation

Kailin Chen, Cheng Peng, Yi Li, Dalong Cheng$^{(\boxtimes)}$, and Si Wei

Computer Vision Group, iFLYTEK Research South China,
Tower 1, Hueng Kong International Innovation Center, No. 37 Jinlong Road,
Guangzhou 511457, China
`dlcheng2@iflytek.com`

Abstract. In order to estimate the spinal curvature, we propose two methods to detect the spinal keypoints at first. In Method-1, we use a RetinaNet to predict the bounding box of each vertebra followed by a HR-Net to refine the keypoint detections. In Method-2, we implement a similar two-stage system, which firstly extract 68 rough points along the spine curves using a Simple Baseline. We then generate patches and make sure each of them contains three vertebrae at most based on ground truth. We train a second Simple Baseline to predict the exact keypoints of these patches, which are generally not fixed in numbers. A delicate postprocess of clustering is proposed to deal with dense keypoint predictions due to the freedom of patch selections. After fusing Method-1 and Method-2, we achieve competitive results on the public leaderboard of AASCE2019 challenge.

1 Introduction

As the current clinical standard for AIS assessment, manual Cobb angle measurement is time-consuming and unreliable, which sparked interest in development of accurate automated spinal curvature estimation in spinal anterior-posterior X-ray images [1–4]. Given the difficulty in using the information of X-rays efficiently, achieving highly accurate automated estimation is rather challenging.

2 Methodology

2.1 Method-1

As shown in Fig. 1, Method-1 is based on RetinaNet [5] and HR-Net [6]. It has two stages. The first stage is to detect vertebrae and to generate corresponding bounding boxes using RetinaNet. We train the RetinaNet using the bounding boxes generated from the 68 keypoints, demanding each box covers 4 keypoints of an individual vertebra. In the second stage, we use HR-Net to detect the 4 keypoints of the bounding box.

K. Chen, C. Peng and Y. Li made equal contribution to the project. D. Cheng constructed and directed the research.

© Springer Nature Switzerland AG 2020
Y. Cai et al. (Eds.): CSI 2019, LNCS 11963, pp. 63–68, 2020.
https://doi.org/10.1007/978-3-030-39752-4_6

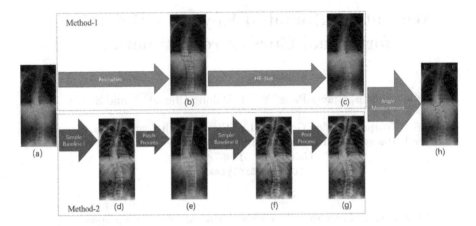

Fig. 1. Framework of the two methods we proposed. (a) is the input image, the blue bounding boxes in (b) and keypoints in (c) are prediction results from RetinaNet and HR-Net in Method-1. The bottom workflow shows the Method-2, including (d) rough points generation, (e) patch selection, (f) patch points generation and (g) clustering. (h) is the final image which shows the process of spine curvature estimation, with the red and blue keypoints respectively representing ground truth and prediction. The lines show which vertebrae are selected to calculate the Cobb angles. (Color figure online)

2.2   Method-2

Following the workflow shown in Fig. 1 Method-2, we firstly train Simple Baseline [7] I with spinal images to detect all 68 keypoints in a spinal sample. In contrast to Method-1 and other similar detection methods, which are difficult to model the scene sequentiality, Simple Baseline I can grasp the global implicit sequentiality of keypoints through generating corresponding heatmaps simultaneously with fixed order. As shown in Fig. 2, such output mechanism avoids the false negatives and false positives in detection methods and leads to a robust model. As a results, the predicted keypoints can smoothly trace the curvature of most spines, even though their landmarks are not accurate enough to calculate Cobb angles. Thus, we use them as the outline sketches of spines to generate patches so as to force the model focus on local information in a certain range of vertebrae.

A patch includes n points, $\{n|n = 4, 6, 8, 10, 12\}$, for one to three vertebrae. Half of the vertebra is allowed, but there must be at least two halves of the vertebra in a patch. Patches are randomly captured in multiple times within a certain vertebrae range. An image sample produces hundreds of patches. Taking into account the fact that the vertebrae vary in permutations and shapes, we believe that the range limitation up to three vertebrae can include most permutations and shapes of vertebrae. Such Patch Process not only makes it much easier to match a template for any adjacent vertebrae, but also makes it easy to augment a large amount of data, so that the following Simple Baseline II becomes a highly robust model.

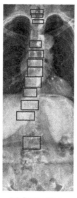 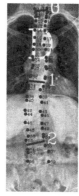 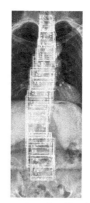

(a) RetinaNet   (b) Simple Baseline I   (c) Patch Process based on Simple Baseline I

Fig. 2. Comparison of local view vertebra detection and global view keypoints detection. (a) are vertebra detection results from RetinaNet and keypoint detection from HR-Net in Method-1. Without order restrictment, detection methods using regression loss are hard to balance the precision and the recall. (b) and (c) are keypoint detection results from Simple Baseline I and Patch Process results in Method-2. With global keypoints detection and sequential scoremap loss, they can cleverly avoid the false negatives and false positives and generate dense patch samples. (Color figure online)

Patches acquired by Patch Process are used to detect keypoints by Simple Baseline II. For each patch, the number of keypoints predicted by Simple Baseline II varies from 4 to 12, corresponding to the patch range from 1 to 3 vertebrae. Given a vertebra is randomly sampled for dozens of times, through Simple Baseline II, a vertebra corresponds to dozens of keypoints with 4 vertex tags, as shown in Fig. 1(f).

2.3 Postprocessing

To summarize the final 68 keypoints from 4 vertex groups of points, we proposed a postprocess to cluster and remove outliers. In Postprocess (i.e. from Fig. 1(f) to (g)), squeezed vertebrae assessment is optional. Therefore, there are 2 results obtained from Method-2.

DBSCAN clustering is applied respectively to 4 vertex groups of points. Clusters within the same groups are sorted in ascending order based on the Y-axis coordinate of cluster centers. The first and the last cluster centers of the 4 vertex groups are then assessed. If the Y-axis coordinate order or the angles of adjacent centers are unnormal, the corresponding clusters are excluded. The following squeezed vertebrea assessment is optional. If a squeezed vertebra exists, the 4 vertexes of it form a right-angled trapezium or parallelogram instead of a isosceles trapezium or rectangle pattern in normal vertebra. Hence, we use the length ratio of two diagonals to remove the redundant points. After the removal, the median number of the remaining cluster centers of the 4 vertex groups is calculated as the number of detected vertebrae. Based on it, K-means clustering is performed. Later, outliers are removed by the gap ratio of the X-axis and Y-axis distance among the adjacent centers. K-means clustering is then performed again. Eventually, the last 3 vertebrae are checked and suppressed if their Y-axis gaps are too small.

3   Experiments and Results

3.1   Implementation Details

In this section, the details about the data processing and how we make the dataset based on the standard dataset will be discussed.

Implementation Details in Method-1. For RetinaNet, we follow [5] to experiment with ResNet-50-FPN backbone. The model is trained for 180k iterations with a total of 8 images per minibatch.

Strategies are proposed for the output of RetinaNet to drop the outlier boxes and keep the remaining boxes stay in the spine line. After bounding box detection, the two IoU of current box with the previous adjacent box and posterior adjacent box will be calculated. If the two IoU values are larger than the threshold T_{IoU}, which is 0.36 in this scheme, the current box will be dropped. The distance between the upper left vertex of current box and previous box is also compared with the distance between the upper left and upper right vertexes of current box, if larger, the current box will also be dropped. These two steps are done through all the bounding boxes in one spine sample.

Following [6], we use HRNet-W48 with input size of 384×288 and the Adam optimizer. The training process is terminated within 30 epochs.

Implementation Details in Method-2. Following [7], we use ResNet-152 backbone with input size of 384×288 for Simple Baseline I. The training ends within 200 epochs with batch size of 32, with CLAHE enhancement.

We choose the Simple Baseline II result from ResNet-152 with input size of 256×192 for further research. The training ends within 70 epochs with batch size of 130.

3.2   Angle Measurement

Different from the usual clinical method [8], we noticed that the angle measurement method provided by organizers is so sensitive and complex that deviation in few pixels can cause great error of angles. So, we tried different angle measurement ways, including fixing the first and the last vertebrae for Proximal-Thoracic (PT) and Thoracic-Lumbar (TL) angles [4], the clinical measurement [8] and the method listed below, which achieved the best result.

The most tilted vertebra above and below the apex are marked as V_1 and V_2. The middle vectors of V_1 and V_2 are used to calculate the intersection angle as the Main Thoracic (MT) Cobb angle. The most tilted vertebra above from V_1 is chosen as V_0. Similarly, the most tilted vertebra below from V_2 is chosen as V_3. PT and TL angle are calculated from the middle vectors of V_0 and V_1, V_2 and V_3, respectively.

3.3   Fusion Method

We chose one result from Method-1 and two results from Method-2 to fuse. The results from Method-2 differ from each other only in the Postprocessing squeezed vertebrea assessment. To combine the results, we consider following fusion methods.

Before voting for the selected vertebrae, the quality of outputs are assessed. If there is a large difference in the length of the opposite sides, it is considered that the current prediction is poor and not under consideration. After quality assessment, the remaining outputs vote for the selected vertebra positions of the separation lines according to the principle of minority subordinate to majority. Once without agreement, one model choice is randomly selected from the patch baseline for the others. In the end, we average over the 3 models of Cobb angles for the final result (Fig. 3).

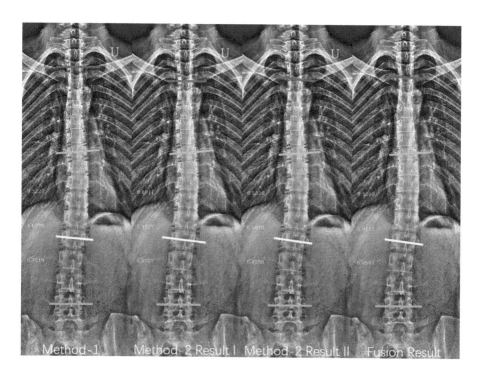

Fig. 3. Results from Method-1, Method-2 and fusion method.

Compared with simply averaging the angles over the schemes, our fusion method decrease 1% SMAPE, reaching 22.16582% eventually.

4   Conclusion

We proposed an automated spinal curvature estimation framework for comprehensive assessment of AIS with two methods.

In Method-1, we implement a RetinaNet for the vertebra bounding box detection, with a HR-Net for local keypoint detection. The vertebra detection based workflow is easy to implement and works efficiently for FOV with unfixed number of vertebrae. However, without sequential modelling, it is hard to balance the precision and the recall.

Method-2 cleverly avoids the disadvantages in Method-1 with a highly robust system, but it needs to design a suitable postprocessing and therefore not easy to implement. In Method-2, we firstly use a Simple Baseline for rough keypoint detection to sketch the spine curve globally. After a well-designed patch generation, we train a second Simple Baseline on patches to predict exact local points with an unfixed number. We implement a delicate clustering postprocessing to handle the prediction of dense keypoints. The highly robust Method-2 grasps the global implicit sequentiality of keypoints and makes it much easier to match a template for any adjacent vertebrae because of various patch samples.

After methods fusion, our framework achieves competitive results in AASCE2019 challenge, which allows clinicians to measure spinal curvature more accurately and robustly.

References

1. Wu, H., Bailey, C., Rasoulinejad, P., Li, S.: Automatic landmark estimation for adolescent idiopathic scoliosis assessment using BoostNet. In: Descoteaux, M., Maier-Hein, L., Franz, A., Jannin, P., Collins, D.L., Duchesne, S. (eds.) MICCAI 2017. LNCS, vol. 10433, pp. 127–135. Springer, Cham (2017). https://doi.org/10.1007/978-3-319-66182-7_15
2. Wang, L., Xu, Q., Leung, S., Chung, J., Chen, B., Li, S.: Accurate automated Cobb angles estimation using multi-view extrapolation net. Med. Image Anal. **58**, 101542 (2019)
3. Chen, B., Xu, Q., Wang, L., Leung, S., Chung, J., Li, S.: An automated and accurate spine curve analysis system. IEEE Access **7**, 124596–124605 (2019)
4. Wu, H., Bailey, C., Rasoulinejad, P., Li, S.: Automated comprehensive Adolescent Idiopathic Scoliosis assessment using MVC-Net. Med. Image Anal. **48**, 1–11 (2018)
5. Lin, T.Y., Goyal, P., Girshick, R., et al.: Focal loss for dense object detection. IEEE Trans. Pattern Anal. Mach. Intell. **PP**(99), 2999–3007 (2017)
6. Sun, K., Xiao, B., Liu, D., et al.: Deep high-resolution representation learning for human pose estimation. Comput. Vis. Pattern Recogn. (2019)
7. Xiao, B., Wu, H., Wei, Y.: Simple baselines for human pose estimation and tracking. Comput. Vis. Pattern Recogn. (2018)
8. Greiner, K.A.: Adolescent idiopathic scoliosis: radiologic decision-making. Am. Fam. Physician **65**(9), 1817–1822 (2002)

Seg4Reg Networks for Automated Spinal Curvature Estimation

Yi Lin[1], Hong-Yu Zhou[2(✉)], Kai Ma[2], Xin Yang[1], and Yefeng Zheng[2]

[1] Department of Electronic Information and Communications,
Huazhong University of Science and Technology, Wuhan, China
[2] Tencent YouTu Lab, Shanghai, China
whuzhouhongyu@gmail.com

Abstract. In this paper, we propose a new pipeline to perform accurate spinal curvature estimation. The framework, named as Seg4Reg, contains two deep neural networks focusing on segmentation and regression, respectively. Based on the results generated by the segmentation model, the regression network directly predicts the cobb angles from segmentation masks. To alleviate the domain shift problem appeared between training and testing sets, we also conduct a domain adaptation module into network structures. Finally, by ensembling the predictions of different models, our method achieves *21.71* SMAPE in the testing set.

Keywords: Spinal curvature estimation · Cobb angle · Deep learning

1 Introduction

Adolescent idiopathic scoliosis (AIS) is the most common form of scoliosis and typically affects children who are at least 10 years old. How to accurately estimate the spinal curvature plays an important role in the treatment planning of AIS. The current clinical standard for AIS assessment relies on doctors' Cobb angle measurement. Such manual intervention process usually makes the operation time-consuming and produces unreliable results. Recently, deep neural networks have got amazing achievements in various image classification tasks. How to apply these deep models to the problem of spinal curvature estimation becomes a hot issue in automated AIS assessment. BoostNet [5] is proposed as a novel framework for automated landmark estimation, which integrates the robust feature extraction capabilities of Convolutional Neural Networks (ConvNet) with statistical methodologies to adapt to the variability in X-ray images. To mitigate the occlusion problem, MVC-Net (Multi-View Correlation Network) [6] and MVE-Net (Multi-View Extrapolation Net) [1] have been developed to make use of features of multi-view X-rays.

Currently, there are two ways to estimate the Cobb angles: (a) predicting landmarks and then angles [5,6] and (b) regressing angle values [1]. The first

---

The first two authors contributed equally.

© Springer Nature Switzerland AG 2020
Y. Cai et al. (Eds.): CSI 2019, LNCS 11963, pp. 69–74, 2020.
https://doi.org/10.1007/978-3-030-39752-4_7

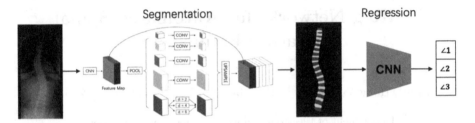

Fig. 1. An overview of our pipeline. We first process the X-ray using a segmentation network. Note that we formalize the groundtruth mask using the provided landmarks. Afterwards, the predicted mask is fed to the regression model to perform angle value prediction.

approach is able to produce high-precision angle results but relies heavily on the landmarks predictions, which means a small mistake in coordinates may lead to a big error in angle predictions. On the contrary, angle regression methods are more stable but may lack the ability to generate precise predictions. In this paper, we explore the possibility of aforementioned two directions in MICCAI AASCE 2019 challenge and our experimental results show that the regression strategy outperforms the landmark approach. We will conduct more details in following sections.

2   Proposed Method

We display our pipeline in Fig. 1. The whole process is constructed by two networks: one for segmentation and the other for regression. The architecture of segmentation network is similar to PSPNet [7] while the regression part employs traditional classification models.

2.1   Preprocessing

We observed that there is an obvious domain gap between training and testing sets (as shown in Fig. 2). To mitigate this problem, we first apply histogram

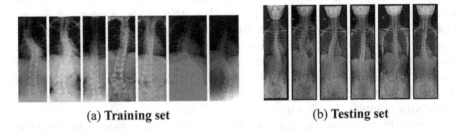

(a) **Training set**   (b) **Testing set**

Fig. 2. A comparison of training set and testing set. It is obvious that these two sets have a huge domain gap.

equalization to both sets to make them visually similar. Considering the limited number of testing images, we decided to manually crop these X-rays to remove the skull and keep the spine in the appropriate scope. Besides, we also applied random rescaling (0.85 to 1.25) and random rotation ($-45°$ to $45°$) during the training process. We tried to add gaussian noise to the input images in order to mitigate the overfitting but it did not work.

For the segmentation task, we built the groundtruth masks on top of offered landmarks' coordinates. It is worth noting that we found adding another class "gap between bones" helped the segmentation model perform the best. We argue that such operation may regularize the training process which makes final predictions more precise.

2.2   Network Architecture

We followed the instructions in [7] to design our segmentation network. After the feature extractor, PSPNet [7] utilized different pooling kernels to capture various receptive fields. In order to keep the feature map size, we also append the dilated convolution with different dilation rates to the pooling pyramid. As shown in Fig. 1, we used 2, 4 and 6 as dilation rates while summing their outputs after the convolution operations. For backbone architecture, we simply took ResNet-50 [4] and ResNet-101 as the basic feature extractor (Fig. 3).

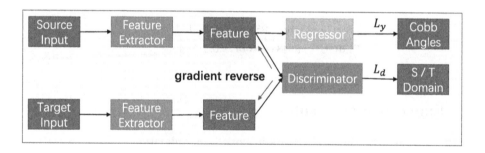

Fig. 3. Our domain adaptation strategy.

As for the classification part, we directly employed recent classification networks to perform the regression task. ImageNet based pretraining was used because we found it helped a lot under limited training samples. Considering the domain gap between training and testing sets, we modified the approach proposed in [3]. The idea is pretty simple which adds a discriminator branch and reverses its gradients during the back propagation so the final loss function can be formalized as:

$$Loss = L_y + \lambda L_d \tag{1}$$

where λ is set to 1 in our experiments.

2.3 Network Training

We used Adam as the default optimizer for both networks where the initial learning rate is 3e−3. β_1 and β_2 are set to 0.9 and 0.999, respectively. We also used weight decay which is 1e−5 and cosine annealing strategy. For the segmentation model, we ran each network for 50 epochs while 90 epochs seem to be a better choice for the regression model. We resized the segmentation input to 1024×512 and regression input to 512×256. The batch size is 32 on 4 NVIDIA P40 GPUs.

Table 1. We made an ablative study on input types and input sizes. It is easy to find that using segmentation mask as input performs the best on the validation set while (512, 256) is the best regression size. The default segmentation backbone is ResNet-50 and regression model is DenseNet-169.

Input type	Input size	Angle1	Angle2	Angle3
Img	(512, 256)	6.0754	7.3386	6.7629
Img + Mask	(512, 256)	5.4489	6.4599	5.8470
Mask	(512, 256)	**4.7128**	**5.7965**	**5.6596**
Mask	(1024, 512)	4.9360	7.2436	6.7321

Table 2. The segmentation performance of PSPNet and DeepLab V3+.

Metric	Ours	PSPNet	DeepLab V3+
mIOU	0.8943	0.8715	0.817

3 Experimental Results

We report our experimental results in both local validation and online testing sets. Note that we did not use cross validation.

3.1 Local Validation

In this part, we report the *L1 distance* between model predictions and groundtruth labels. As shown in Table 1, it is obvious that segmentation mask is the best input type and (512, 256) is the best input size. In Table 2, we compare the performance of our improved version with PSPNet and DeepLab V3+ [2]. We can find that adding a dilation pyramid thus improves the performance of previous PSPNet. It is interesting that PSPNet surpasses DeepLab V3+ by a large margin since they have achieved comparable performance in PASCAL VOC segmentation task. We argue that the failure of DeepLab can be attributed to the limited training data and our parameter tuning strategies.

3.2 Online Testing

We formalize our online testing results into Table 3. We can see that our dilation pyramid improves the online SMAPE by 0.48. Also, it is quite normal to see that EfficientNet-b5 is better than DenseNet-169 considering its higher ImageNet performance. After adding domain adaptation module, the single model performance rises to 26.15. During the model ensemble stage, we assigned different weights to different model outputs considering their validation scores. We mainly ensembled ResNet series, DenseNet series and EfficientNet series. This strategy helped us to improve the SMAPE score to 22.25.

Table 3. The segmentation performance of PSPNet and DeepLab V3+.

Strategies					
PSPNet + DenseNet-169	✓				
Ours + DenseNet-169		✓			
Ours + EfficientNet-b5			✓	✓	
Domain Adaptation				✓	✓
Model Ensemble					✓
SMAPE	28.51	28.03	27.07	26.15	22.25

Angles, Training Data

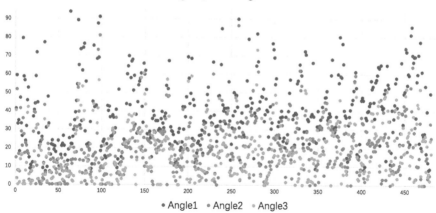

Fig. 4. Distribution of 3 angles in the training set.

During the online testing stage, we revisited the distribution of 3 angles in the training set. From Fig. 4, we can easily find out that Angle2 and Angle3 are much smaller than Angle2. Also, Angle2 has many values which are close to zero. According to such phenomenons, we reduced angles smaller than 4° to zeros which brought us to 21.71 SMAPE.

References

1. Chen, B., Xu, Q., Wang, L., Leung, S., Chung, J., Li, S.: An automated and accurate spine curve analysis system. IEEE Access (2019)
2. Chen, L.C., Zhu, Y., Papandreou, G., Schroff, F., Adam, H.: Encoder-decoder with atrous separable convolution for semantic image segmentation. In: Proceedings of the European Conference on Computer Vision (ECCV), pp. 801–818 (2018)
3. Ganin, Y., Lempitsky, V.: Unsupervised domain adaptation by backpropagation. arXiv preprint arXiv:1409.7495 (2014)
4. He, K., Zhang, X., Ren, S., Sun, J.: Deep residual learning for image recognition. In: CVPR, pp. 770–778 (2016)
5. Wu, H., Bailey, C., Rasoulinejad, P., Li, S.: Automatic landmark estimation for adolescent idiopathic scoliosis assessment using BoostNet. In: Descoteaux, M., Maier-Hein, L., Franz, A., Jannin, P., Collins, D.L., Duchesne, S. (eds.) MICCAI 2017. LNCS, vol. 10433, pp. 127–135. Springer, Cham (2017). https://doi.org/10.1007/978-3-319-66182-7_15
6. Wu, H., Bailey, C., Rasoulinejad, P., Li, S.: Automated comprehensive Adolescent Idiopathic Scoliosis assessment using MVC-Net. Med. Image Anal. **48**, 1–11 (2018)
7. Zhao, H., Shi, J., Qi, X., Wang, X., Jia, J.: Pyramid scene parsing network. In: Proceedings of the IEEE Conference on Computer Vision and Pattern Recognition, pp. 2881–2890 (2017)

Automatic Spine Curvature Estimation by a Top-Down Approach

Shixuan Zhao, Bo Wang, Kaifu Yang, and Yongjie Li[✉]

MOE Key Lab for Neuroinformation, School of Life Science and Technology,
University of Electronic Science and Technology of China, Chengdu 610054, China
liyj@uestc.edu.cn
http://www.neuro.uestc.edu.cn/vccl/home.html

Abstract. Adolescent idiopathic scoliosis (AIS) is the most common type of spinal deformity in children in early puberty. Cobb angle is widely used in the diagnosis and treatment of scoliosis. However, existing Cobb angle measurement methods in clinical practice are time-consuming and unreliable. Accurate quantitative assessment of spinal curvature is an essential task in the clinical assessment and treatment planning of AIS. In this study, we proposed a top-down approach to accomplish the task of automatic spinal curvature estimation. We achieved 26.0535% of symmetric mean absolute percentage error (SMAPE).

Keywords: AIS · Deep learning · Cobb angles · Spinal curvature

1 Introduction

Adolescent idiopathic scoliosis (AIS) is the most common type of spinal deformity in children in early puberty [10]. It dramatically affects the quality of life, which can cause complications from heart and lung injuries in severe cases. Currently, clinicians make treatment decisions by assessing the extent of spinal deformity. The criteria for assessing scoliosis is the manual measurement of Cobb angles [5] based on x-ray images of anterior-posterior (AP) and lateral (LAT), which are widely used in the diagnosis and treatment strategies for scoliosis. However, existing Cobb angle measurement methods in clinical practice are time-consuming and unreliable [6,8]. Therefore, accurate computer-assisted spinal curvature measurement is necessary.

Computer-assisted quantitative evaluation of spinal curvature has been increasingly studied recently. Anitha et al. [2,3] used active contouring and filtering to locate the desired vertebrae and segment them, then calculated the Cobb angle based on the segmentation results. Li et al. [11] estimated vertebral coordinates using BoostNet based on deep learning. Wu et al. [12] combined the multi-view X-ray features with a multi-view convolutional layer to construct a Multi-View Correlation Network (MVC-Net) to estimate the Cobb angle. Wang et al. [9] directly predicted the Cobb angle using a Multi-View Extrapolation Net

© Springer Nature Switzerland AG 2020
Y. Cai et al. (Eds.): CSI 2019, LNCS 11963, pp. 75–80, 2020.
https://doi.org/10.1007/978-3-030-39752-4_8

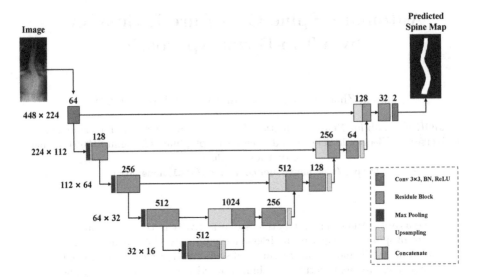

Fig. 1. The architecture of the RU-Net for the spinal segmentation. The arrows represent the connection of the data. A colored box corresponds to the operation of the feature map. (Color figure online)

(MVE-Net). Chen et al. [4] used an Adaptive Error Correction Net (AEC-Net) to convert the estimation of spinal Cobb into a high-precision regression task.

Presently, AASCE 2019 sets up a challenge [1]. The purpose of the challenge is to perform accurate automatic spinal curvature estimation from the spine AP X-ray images, which aims to study the (semi)automatic spine curvature estimation algorithm. We proposed a top-down approach to estimate and analyze the spine curvature.

2   Methods and Results

Due to the high complexity and variability of spine X-rays, it is difficult to achieve high accuracy if deep learning is used directly for end-to-end learning. In this paper, we proposed a top-down approach to accomplish the task of automatic spinal curvature estimation. We first segmented the spine region through a spinal segmentation network. Then, the segmentation result is used to fit the trend of the spine. Finally, the Cobb angles were obtained after the adaptive vertebral angle correction performed at the maximum degree of curvature.

2.1   Spine Segmentation Network

We use RU-Net to segment the spine, which is improved using U-Net [7], as shown in Fig. 1. We replace the convolution module of the original network in the residual module to enhance the use of context information. The downsampling part is pooled using a 2 × 2 max pooling. The upsampling portion is

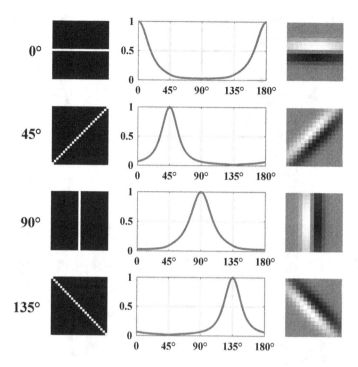

Fig. 2. The left column is the edge map in different directions as input. The middle graphs are the corresponding function of the Gaussian kernel for the input in different directions. The pictures on the right are the appearance of the corresponding most massive Gaussian kernel. (Color figure online)

replaced by a bilinear difference layer instead of a deconvolution layer, which further reduces the parameters of the network to prevent over-fitting. Finally, the network outputs two layers of feature maps to predict the spine and background separately. The cross-entropy loss is calculated after passing through the softmax layer.

2.2 Algorithm for Predicting the Trend of the Spine

We find that the segmentation results can reflect the degree of curvature of the spine to a certain extent. We assume that the curved spine like a soft water pipe. The boundaries of the spine are like the wall of a water pipe, and the degree of bending of the flowing water in it is the target we need to quantify. First, we use the segmentation results to find the left and right boundaries of the spine. Then we search the nearest neighbor to the right boundary and connect it to each point on the left boundary. Similarly, we also get a series of midpoints in the opposite direction. The midpoint of the line is the water flow. Finally, we fit the points to the function of spine simulation by the least-squares method, which completes the quantification of the spine, as shown in Fig. 3(a–d) middle.

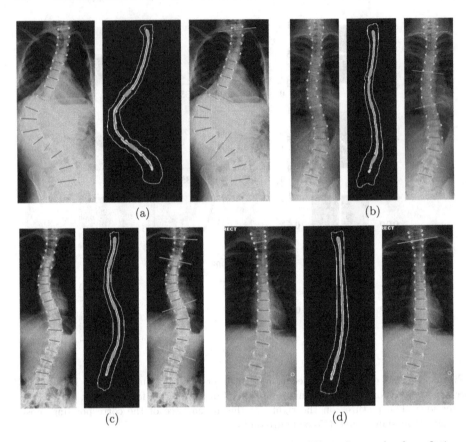

Fig. 3. The pictures on the left are ground truth. The middle is the result of our fitting on the spine. The images on the right are the result of vertebral angle correction. (Color figure online)

We use this function to calculate the position that has the maximal curvature of the spine from which we can directly obtain the Cobb angles.

2.3   Vertebrae Angle Correction

Our function can only get the general trend of the spine instead of simulating the condition of each vertebra since the spine is not continuous, but the vertebrae and intervertebral discs are combined. In response to this problem, we conduct post-processing. Firstly, based on the results obtained in the previous step, we obtained the edge map of the vertebrae by edge extraction using the canny operator at the position where the curvature of the spine is largest. We assume that the outline of the vertebra is significantly larger than the other small edges, and the most significant direction of the edge gradient should be perpendicular to the direction of the vertebrae. Therefore, we construct a flat Gaussian kernel

(Fig. 2), which can rotate 180°. We use the directional derivative of this Gaussian kernel and the edge map of the vertebrae to obtain a response curve in different directions, which is the direction of the most significant response in the direction of the vertebrae. The Cobb angles are calculated using the direction of the vertebrae after correction.

2.4   Result

Figure 3 shows the spine AP x-ray images of four patients. Left is the original map, the red line is the central axis of each vertebra perpendicular to the direction of the spine, and the Cobb angle is the angle between the green segments. The result of the segmentation of the spine is shown in the white line of the middle image. The yellow dots represent the trend of the spine, and the red curve is the fitted spine function obtained by the least-squares method. The green square is the position that has the max curvature of the spine, and it can be found to be in excellent agreement with the ground truth. Right pictures compare the results of our models and the ground truth, and the cyan line is the direction in which the vertebrae angle is corrected. After the correction, the direction of the vertebrae is almost the same as the result of the doctor.

Acknowledgements. This work was supported by Natural Science Foundations of China under Grant 61703075 and 61806041, Sichuan Province Science and Technology Support Project under Grant 2017SZDZX0019.

References

1. Accurate automated spinal curvature estimation MICCAI 2019 challenge (2019). https://aasce19.grand-challenge.org/Home/
2. Anitha, H., Karunakar, A., Dinesh, K.: Automatic extraction of vertebral endplates from scoliotic radiographs using customized filter. Biomed. Eng. Lett. **4**(2), 158–165 (2014)
3. Anitha, H., Prabhu, G.: Automatic quantification of spinal curvature in scoliotic radiograph using image processing. J. Med. Syst. **36**(3), 1943–1951 (2012)
4. Chen, B., Xu, Q., Wang, L., Leung, S., Chung, J., Li, S.: An automated and accurate spine curve analysis system. IEEE Access **7**, 124596–124605 (2019)
5. Cobb, J.: Outline for the study of scoliosis. Instr. Course Lect. AAOS **5**, 261–275 (1948)
6. Pruijs, J., Hageman, M., Keessen, W., Van Der Meer, R., Van Wieringen, J.: Variation in cobb angle measurements in scoliosis. Skeletal Radiol. **23**(7), 517–520 (1994)
7. Ronneberger, O., Fischer, P., Brox, T.: U-net: convolutional networks for biomedical image segmentation. In: Navab, N., Hornegger, J., Wells, W.M., Frangi, A.F. (eds.) MICCAI 2015. LNCS, vol. 9351, pp. 234–241. Springer, Cham (2015). https://doi.org/10.1007/978-3-319-24574-4_28
8. Vrtovec, T., Pernuš, F., Likar, B.: A review of methods for quantitative evaluation of spinal curvature. Eur. Spine J. **18**(5), 593–607 (2009)

9. Wang, L., Xu, Q., Leung, S., Chung, J., Chen, B., Li, S.: Accurate automated cobb angles estimation using multi-view extrapolation net. Med. Image Anal. **58**, 101542 (2019)
10. Weinstein, S.L., Dolan, L.A., Cheng, J.C., Danielsson, A., Morcuende, J.A.: Adolescent idiopathic scoliosis. Lancet **371**(9623), 1527–1537 (2008)
11. Wu, H., Bailey, C., Rasoulinejad, P., Li, S.: Automatic landmark estimation for adolescent idiopathic scoliosis assessment using BoostNet. In: Descoteaux, M., Maier-Hein, L., Franz, A., Jannin, P., Collins, D.L., Duchesne, S. (eds.) MICCAI 2017. LNCS, vol. 10433, pp. 127–135. Springer, Cham (2017). https://doi.org/10.1007/978-3-319-66182-7_15
12. Wu, H., Bailey, C., Rasoulinejad, P., Li, S.: Automated comprehensive adolescent idiopathic scoliosis assessment using MVC-net. Med. Image Anal. **48**, 1–11 (2018)

Automatic Cobb Angle Detection Using Vertebra Detector and Vertebra Corners Regression

Bidur Khanal[1(✉)], Lavsen Dahal[1], Prashant Adhikari[2], and Bishesh Khanal[1]

[1] NepAl Applied Mathematics and Informatics Institute for Research (NAAMII),
Suryabinayak, Nepal
{bidur.khanal,lavsen.dahal,bishesh.khanal}@naamii.org.np
[2] Hospital for Advanced Medical Surgery (HAMS), Kathmandu, Nepal

Abstract. Correct evaluation and treatment of Scoliosis require accurate estimation of spinal curvature. Current gold standard is to manually estimate Cobb Angles in spinal X-ray images which is time consuming and has high inter-rater variability. We propose an automatic method with a novel framework that first detects vertebrae as objects followed by a landmark detector that estimates the 4 landmark corners of each vertebra separately. Cobb Angles are calculated using the slope of each vertebra obtained from the predicted landmarks. For inference on test data, we perform pre and post processings that include cropping, outlier rejection and smoothing of the predicted landmarks. The results were assessed in AASCE MICCAI challenge 2019 which showed a promise with a SMAPE score of 25.69 on the challenge test set.

Keywords: Scoliosis · Landmark · Object detection · Cobb angle

1 Introduction

Scoliosis is a sideways curvature of the spine occurring mostly in teens. Severe scoliosis can also lead to disability. The current gold standard for diagnosing scoliosis is manual measurement of Cobb Angles in anterior-posterior (AP) or lateral (LAT) X-ray images which involve identifying the most tilted vertebrae above and below the apex of the spinal curve [1]. However, the procedure is time-consuming and observer dependent, leading to high inter-observer variability that could negatively impact assessing prognosis and treatment decisions [2]. Thus, there has been increasing interest in automatic estimation of Cobb angles directly from the X-ray images. In this context, we participated in MICCAI 2019 challenge on Accurate Automated Spinal Curvature Estimation (AASCE)[1] where the task was to accurately estimate three Cobb angles [3] from the training dataset containing 609 AP x-rays[2] whose results were assessed on 98 test images. The ground truth (GT) annotations are the anatomical landmarks consisting of four corners of 17 vertebrae: twelve thoracic and five lumbar.

---

[1] https://aasce19.grand-challenge.org/Home/.

[2] http://spineweb.digitalimaginggroup.ca/spineweb/index.php?n=Main.Datasets.

© Springer Nature Switzerland AG 2020
Y. Cai et al. (Eds.): CSI 2019, LNCS 11963, pp. 81–87, 2020.
https://doi.org/10.1007/978-3-030-39752-4_9

Related Work: The two most common approaches of estimating Cobb angles are Segmentation based and Landmark based approaches. The segmentation based methods first segment all the vertebrae or the end-plates of the vertebrae to identify the most tilted vertebrae from which the Cobb angles are estimated [3–5]. Accurate segmentation of each vertebra from X-ray images is difficult with traditional feature-engineering based approaches. To our knowledge, even modern supervised deep neural networks are not robust and accurate enough yet for the vertebra segmentation. Creating accurate GT segmentation is time consuming and relatively difficult compared to annotating landmarks: four corners of the vertebrae. In Landmark based approach which is the state-of-the-art, the four corners of each vertebrae are detected and are subsequently used for estimating Cobb angles. Some methods jointly estimate all the landmarks and Cobb angles, while others first estimate landmarks followed by Cobb angle computation which might include outlier rejection and post-processing techniques [6,7].

There are several approaches of detecting landmarks in medical images such as Reinforcement learning [8], iterative patch based approaches [9] and fully convolutional neural network based approaches [10]. One important difference in vertebra landmarks compared to other anatomical landmarks is the presence of a large number of similar looking vertebrae. We believe that detecting vertebrae as objects before finding landmarks within the detected vertebrae is advantageous as it allows: (i) avoiding difficulty for translation equivariant CNNs to learn very different coordinate locations for almost identical appearing vertebrae (ii) leveraging popular object detectors pre-trained for computer vision tasks (iii) Reducing the search space for landmark detector.

Contribution: We propose a novel approach to first detect 17 vertebrae with a bounding box object detector, after which each of the predicted boxes is fed to a landmark detector as illustrated in Fig. 1. The predicted landmarks are post-processed to remove outliers before calculating the three Cobb angles. [11] used Faster-RCNN [12] object detector to detect intervertebral disc in lateral X-rays, but they left the landmark detection as a future work.

2   Dataset

The dataset consists of 609 spinal AP x-ray images available at SpineWeb[3] as *Dataset 16*. Each image has 68 GT landmarks corresponding to 4 corners of the 17 vertebrae, and 3 Cobb Angles. Organizers provided test images without GT separately. We connected the four landmark corners of each vertebrae to create a box whose width and height were then increased symmetrically by 50 and 10 pixels respectively to create GT bounding boxes. All the bounding boxes were labelled as belonging to a single class. The GT bounding boxes were used to crop and extract individual vertebrae as a single separate image containing four landmark corners. The coordinates of the landmarks are normalized to the coordinate system that maps all the pixel coordinates of the cropped image to

---

[3] http://spineweb.digitalimaginggroup.ca/spineweb/index.php?n=Main.Datasets.

the interval [0, 1]. The normalized landmark coordinates are used as GT labels for the landmark regression network.

3   Vertebrae Detection Followed by Landmarks Regression

We use an object detector to detect the vertebrae as bounding box objects which are then fed to a landmark regression network as separate input images. The predicted normalized landmark coordinates from individual bounding boxes are combined and mapped back to the original images as shown in Fig. 1.

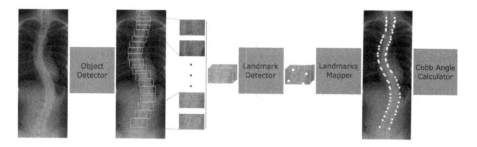

Fig. 1. Proposed Framework: The input images are first passed to an object detector that detects the vertebrae. The detected vertebrae are extracted as individual images and passed to a landmark detector that detects the four corners of the vertebra. The landmarks are mapped back to the original image from which Cobb angles are calculated. We use CNN-based Faster-RCNN and DenseNet for object detection and landmark detection respectively.

3.1   Training Vertebra Detection with Faster-RCNN

Faster-RCNN [12] is a widely used two-stage object detector consisting of: (i) a Region Proposal Network (RPN) that proposes potential object regions from a set of anchor boxes of various sizes in a sliding window over the feature maps extracted from a CNN-based base network (ii) a fully connected and a bounding box regression layer that regress bounding box locations of the identified objects. We used ResNet V1 101 with pre-trained weights on Imagenet data[4] as the base network, which was fine-tuned after block 2. We used two scales with box areas of 64^2 and 128^2 pixels, and aspect ratios 1:1 and 2:1 for RPN's anchor boxes, as the vertebrae are relatively small and do not have extreme aspect ratios. The network was trained for around 180k steps with batch size 1 using SGD optimizer with momentum 0.9, learning rate 0.0003 and early stopping. The implementation was adopted from Luminoth[5] in Tensorflow framework 10.1. Data augmentation

---

[4] https://github.com/tensorflow/models/tree/master/research/slim.
[5] https://github.com/tryolabs/luminoth.

included random Gaussian noise ($\mu = 0$, $\sigma = 0.005$), and vertical and horizontal flips with a probability of 0.5. All the images were rescaled preserving the aspect ratio such that its sizes remained within 600–1000 pixels as much as possible.

3.2   Training Landmark Detector with DenseNet

The four corner landmarks were estimated using a Densely Connected Convolutional Neural Network (DenseNet) which are known to require fewer parameters than traditional CNN [13]. In DenseNet, each layer's feature maps are used for all subsequent layers within a block, where each block constitutes a bottleneck layer (a 2d Convolution layer with 1×1 filter size), batch normalization, ReLU activation, and a regular 2D convolution layer (3×3 filter size). We used 5 blocks with a growth rate of 8 which is the number of output feature maps of each layer. The 2D Global Average Pooling is used after 5 blocks followed by a dense layer. The final layer consists of 8 output units with a linear activation function. All the input images to landmark detector were resized to 200×120 pixels.

4   Pre and Post Processing During Inference

Cropping: Almost all test images contained skull and pelvic regions but none of the training images had them. During training, the model did not see negative samples of skull and pelvic regions making it prone to falsely detect structures appearing similar to vertebra such as jaws. We randomly picked one test image with an aspect ratio a_0 and found empirically that cropping $c_{t_0} = 0.18$ and $c_{b_0} = 0.21$ times the image height from the top and bottom removed skull and pelvic regions satisfactorily. All the remaining test images with aspect ratio a were cropped by $c_t = c_{t_0} \cdot \frac{a}{a_0}$ and $c_b = c_{b_0} \cdot \frac{a}{a_0}$ fraction of the image height from the top and bottom respectively.

Outlier Rejection: We removed some of the outliers by using the fact that adjacent vertebrae cannot be far away from each other: if the x-center (horizontal) of any detected bounding box is more than half box width away from the x-centers of both of its two nearest neighboring (top and bottom) boxes, they are rejected as outliers. For the topmost and bottom boxes, the same test was done against only one nearest neighbor.

Curve Fitting and Cobb Angle Calculation from Predicted Landmarks: We used the code provided along with the challenge dataset [6] to calculate 3 Cobb angles - Main Thoracic (MT), Proximal Thoracic (PT) and Thoracolumbar/Lumbar (TL/L) from a given set of landmarks. It did not work well when the number of landmarks were not exactly 68 corresponding to the 17 bounding boxes. To ensure exactly 68 landmark points for angle calculation, we used the following after outlier rejection: when the detected vertebrae number is more than 17, reject extra bounding boxes starting from the bottom. Similarly, if the number is less than 17, duplicate the bottom landmarks as required. We also smoothed the landmarks by fitting a polynomial curve where the degree 6 polynomial gave best fit out of 3 to 8 on visual inspection. The x-coordinate of each landmark is regressed by using the y-coordinate as the independent variable of the fitted polynomial. The smoothed landmarks were the ones that were used to estimate Cobb angles in the final test score.

5 Results

The results were evaluated with symmetric mean absolute percentage error, SMAPE$= \frac{1}{N} \sum_N \frac{\sum_m |a_g - a_p|}{\sum_m (a_g + a_p)} 100\%$, where we have $N (= 98)$ test images, $m (= 3)$ Cobb angles per image, GT angle a_g and the corresponding predicted angle a_p.

Table 1. Results for different experiment setups

Exp no.	Processing for test images	SMAPE
1	No Cropping	33.3%
2	Cropping and outlier removal without smoothing	26.79%
3	Cropping, outlier removal and smoothing with order 6 polynomial fitting	25.69%

Table 1 shows the results of three different experiments where we achieved our best score in the challenge by cropping, rejecting outliers and smoothing the estimated landmarks. The top score in the leader board was 21.71% when the challenge results entry was closed. Figure 2 shows detected bounding boxes and landmarks, and results of outlier rejection and smoothing with polynomial fitting in 4 example images from test set.

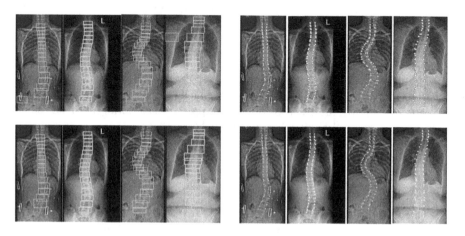

Fig. 2. Anti-clockwise from top left: bounding box detection, outlier rejection, landmark prediction and smoothing of landmarks (green) with polynomial degree 6 for four images of the test set (Color figure online)

6   Discussion and Conclusion

Detecting vertebrae as objects before predicting corner landmarks is found to be a promising approach. However, cropping all test images will not generalize well. A more robust object detector trained with images having negative samples from skull and pelvic regions could eliminate the need of cropping. The proposed approach does not properly take into account the inter-dependency between landmark positions of different vertebrae. A learning algorithm to learn this inter-dependency could improve the results. Finally, learning to estimate the angles directly from landmarks instead of using the geometric algorithm could be robust to noisy landmark prediction.

Acknowledgements. This work is supported by NVIDIA GPU donation. We also thank Pro-Mech Minds & Engineering Services for agreeing to partially fund conference visit expenses for presenting this work.

References

1. Greiner, K.A.: Adolescent idiopathic scoliosis: radiologic decision-making. Am. Fam. Physician **65**, 1817–22 (2002)
2. Loder, R.T., et al.: The assessment of intraobserver and interobserver error in the measurement of noncongenital scoliosis in children $\leq$ 10 years of age. Spine **29**(22), 2548–2553 (2004)
3. Sardjono, T.A., et al.: Automatic cobb angle determination from radiographic images. Spine (Phila Pa 1976) **38**, E1256–E1262 (2013)
4. Allen, S., et al.: Validity and reliability of active shape models for the estimation of cobb angle in patients with adolescent idiopathic scoliosis. J. Digit. Imaging **21**, 208–18 (2008)
5. Zhang, J., et al.: Automatic cobb measurement of scoliosis based on fuzzy hough transform with vertebral shape prior. J. Digit. Imaging **22**, 463–72 (2009)
6. Wu, H., Bailey, C., Rasoulinejad, P., Li, S.: Automatic landmark estimation for adolescent idiopathic scoliosis assessment using boostnet. In: Descoteaux, M., Maier-Hein, L., Franz, A., Jannin, P., Collins, D.L., Duchesne, S. (eds.) MICCAI 2017. LNCS, vol. 10433, pp. 127–135. Springer, Cham (2017). https://doi.org/10.1007/978-3-319-66182-7_15
7. Sun, H., Zhen, X., Bailey, C., Rasoulinejad, P., Yin, Y., Li, S.: Direct estimation of spinal cobb angles by structured multi-output regression. In: Niethammer, M., et al. (eds.) IPMI 2017. LNCS, vol. 10265, pp. 529–540. Springer, Cham (2017). https://doi.org/10.1007/978-3-319-59050-9_42
8. Alansary, A., et al.: Evaluating reinforcement learning agents for anatomical landmark detection. MedIA **53**, 156–164 (2019)
9. Li, Y., et al.: Fast multiple landmark localisation using a patch-based iterative network. In: Frangi, A.F., Schnabel, J.A., Davatzikos, C., Alberola-López, C., Fichtinger, G. (eds.) MICCAI 2018. LNCS, vol. 11070, pp. 563–571. Springer, Cham (2018). https://doi.org/10.1007/978-3-030-00928-1_64
10. Payer, C., et al.: Integrating spatial configuration into heatmap regression based CNNS for landmark localization. MedIA **54**, 207–219 (2019)

11. Sa, R., et al.: Intervertebral disc detection in X-ray images using faster R-CNN. In: Conference Proceedings of IEEE Engineering in Medicine and Biology Society, pp. 564–567 (2017)
12. Ren, S., et al.: Faster R-CNN: towards real-time object detection with region proposal networks. IEEE TPAMI **39**(6), 1137–1149 (2017)
13. Huang, G., et al.: Densely connected convolutional networks. In: CVPR (2017)

Automated Estimation of the Spinal Curvature via Spine Centerline Extraction with Ensembles of Cascaded Neural Networks

Florian Dubost[1(✉)], Benjamin Collery[1,2], Antonin Renaudier[1,2], Axel Roc[1,2], Nicolas Posocco[1,3], Wiro Niessen[1,4], and Marleen de Bruijne[1,5]

[1] Department of Radiology and Nuclear Medicine,
Erasmus University Medical Center, Rotterdam, The Netherlands
floriandubost1@gmail.com, marleen.debruijne@erasmusmc.nl
[2] Ecole des Mines de Saint-Etienne, Saint-Etienne, France
[3] Ecole Centrale Marseille, Marseille, France
[4] Department of Imaging Physics, Faculty of Applied Science, TU Delft,
Delft, The Netherlands
[5] Department of Computer Science, University of Copenhagen,
Copenhagen, Denmark

Abstract. Scoliosis is a condition defined by an abnormal spinal curvature. For diagnosis and treatment planning of scoliosis, spinal curvature can be estimated using Cobb angles. We propose an automated method for the estimation of Cobb angles from X-ray scans. First, the centerline of the spine was segmented using a cascade of two convolutional neural networks. After smoothing the centerline, Cobb angles were automatically estimated using the derivative of the centerline. We evaluated the results using the mean absolute error and the average symmetric mean absolute percentage error between the manual assessment by experts and the automated predictions. For optimization, we used 609 X-ray scans from the London Health Sciences Center, and for evaluation, we participated in the international challenge "Accurate Automated Spinal Curvature Estimation, MICCAI 2019" (100 scans). On the challenge's test set, we obtained an average symmetric mean absolute percentage error of 22.96.

1 Introduction

Diagnosis and treatment planning of Adolescent idiopathic scoliosis (AIS) relies on the estimation of the spinal curvature, which can be measured using Cobb angles [2]. We propose an automated method to measure Cobb angles in X-ray scans. In our approach, the centerline of the spine was automatically segmented using cascaded neural networks, that were optimized end-to-end, i.e. trained

---

B. Collery, A. Renaudier and A. Roc—equal contribution.

© Springer Nature Switzerland AG 2020
Y. Cai et al. (Eds.): CSI 2019, LNCS 11963, pp. 88–94, 2020.
https://doi.org/10.1007/978-3-030-39752-4_10

simultaneously. While the first network focused on the segmentation of the spine, the second network focused on the extraction of the centerline of the spine, using the results of the first network. The centerline was then smoothed and its derivative was computed to obtain tangents and estimate the Cobb angles.

Unlike our approach, most automated methods for estimating Cobb angles rely on the segmentation of the individual vertebrae or the detection of their corners [3, 4, 8]. Recently, Wu et al. [3] predicted the position of landmarks indicating the corners of vertebrae using convolution neural networks (CNNs). The authors proposed BoostNet, a layer architecture to reduce intra-class variance of feature embeddings. They also designed a structured output for their networks, which incorporates respective positions of landmarks in a connectivity matrix. Wu et al. [3] achieved very accurate results in the automated localization of vertebrae landmarks, but did not measure Cobb angles in their paper. Horng et al. [4] recently proposed to measure the spinal curvature by first isolating spine region and detecting each vertebra using image processing, subsequently segmenting vertebrae with a U-net like network [1], and measuring Cobb angles using the vertebrae segmentations. Similarly to our approach, Okashi et al. [9] measured Cobb angles directly from the centerline of the spine. In their work, the centerline was extracted using a complex hand-engineered image processing algorithm.

2 Methods

2.1 Datasets

The method was optimized with a publicly available dataset of 609 spinal anterior-posterior X-ray images [3] with 17 vertebrae of the thoracic and lumbar regions manually indicated using landmarks at the four corners thereof. Cobb angles were computed using the landmarks. To evaluate our method, we competed in the Accurate Automated Spinal Curvature Estimation (AASCE) challenge[1] hosted by the 22nd International Conference on Medical Image Computing and Computer Assisted Intervention. The test dataset contained 100 X-ray scans. These images were different from the training images, which focused on the spine, in that they displayed areas of the neck and shoulders. We manually cropped the images of the evaluation dataset so that they show the same region as images of the training dataset. In addition, this dataset was substantially different from the training dataset in that the majority of its images had small Cobb angles (low curvature), while Cobb angles in the training dataset were more uniformly distributed. As additional preprocessing, we normalized the intensities of all images by dividing by the maximum intensity value of each image individually.

As ground truths, the challenge organisers provided three Cobb angles that were manually measured: the *major Cobb angle* which estimates the highest overall spinal curvature (the Cobb angle usually reported in the litterature);

---

[1] https://aasce19.grand-challenge.org/Home/.

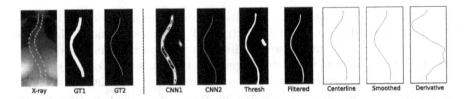

Fig. 1. Spine segmentation and curvature computation. Top row, left to right: an input X-ray scan with corners of the vertebrae manually indicated in blue (*X-ray*); spine segmentation by connecting vertebral corners (*GT1*) and centerline (*GT2*), used for training the cascaded networks. *CNN1* and *CNN2* are the outputs of the cascaded networks: the spine and spine centerline segmentations, respectively. The bottom row illustrates the post-processing pipeline: thresholding the centerline segmentation (*Thresh*); removing small connected components (*Filtered*); extracting the spine centerline curve (*Centerline*); centerline smoothing using heat equation (*Smoothed*); computing the derivative of the centerline (*Derivative*), which was subsequently used to compute Cobb angles.

the *upper Cobb angle*, which is the Cobb angle measuring the highest spinal curvature in the spine region above the region of highest overall curvature; and the *lower Cobb angle*, which is the Cobb angle in the spine region under the region of highest curvature.

2.2   Centerline Extraction with Neural Networks

Two cascaded convolutional neural networks were used to extract the centerline of the spine. Their architecture was that of a U-Net [1] with fewer feature maps and batch normalization layers [5] before each pooling layer (Fig. 2). The networks were optimized end-to-end, simultaneously. The first network was given the X-ray scan as input and was optimized to segment the complete spine, while the second network was given the output of the first network as input and was optimized to segment only the centerline. In earlier experiments, we used a single network to segment the centerline directly, which led to parts of the centerline – often areas with low contrast – not being segmented. Ground truth segmentations of the complete spine and the centerline were automatically computed using the landmarks provided in the training dataset.

Fig. 2. Cascaded networks architecture. The last number in each layer is the number of feature maps.

The networks were optimized using Adadelta optimizer [6] and mean squared error loss function over voxel intensities between the binary ground truth segmentations of the spine and centerline and the segmentations predicted by the network. Images were augmented online during training, with random rotation, translation, and horizontal flipping, addition of Gaussian noise, and varying brightness and contrast.

2.3 Postprocessing

Curve Extraction. The network output was binarized using a low threshold (0.25) to ensure continuity of the centerline, followed by small (smaller than 40×140 pixels) connected components removal. Borders of the centerline were detected to compute two curves. Points in the middle of both curves were then selected to model the centerline curve. In our experiments on the validation set, this was the most robust approach to extract the curve in the training dataset.

Smoothing with Heat Equation. Centerline curves extracted from the images showed some local noise due to image resolution and errors in the segmentation. To compute more accurate derivatives, the centerline was smoothed using the heat equation, solved with Euler method. There were two parameters: the heat transfer coefficient, which was set to 0.01 and the number of iterations in Euler's method set to 65000, which had to be large enough and was tuned on the training set. We did not experiment with varying the heat transfer coefficient.

Computation of Cobb Angles. Similarly to Horng et al. [4], we computed Cobb angles as

$$\phi = \frac{180}{\pi} \left| arctan \left(\frac{T(p_{R,M}) - T(p_{R,m})}{1 + T(p_{R,M}) \cdot T(p_{R,m})} \right) \right|, \tag{1}$$

where $T(p)$ is the tangent slope of the centerline at point p, $p_{R,M}$ is the point with the maximum slope in region R, and $p_{R,m}$ is the point with the minimum slope in R. We computed the derivative of the centerline to obtain the tangent of the centerline in every point. Then, we selected 19 points evenly spaced over the derivative of the centerline ("Derivative" in Fig. 1) as an approximation of the location of the interstices between the 17 annotated vertebrae, and considered values of the slope of the tangent, $T(p)$, only in those points. Considering tangents over the entire curve would quantify its curvature more accurately but would be less comparable to the manually measured Cobb angles.

To compute the major angle, the entire centerline was used as the region of interest R. The upper and lower angles were computed in the regions above and below the region of the highest curvature, defined as the region between $p_{R,m}$ and $p_{R,M}$.

2.4 Ensemble of Ensembles

To improve the generalization performance of our models, we created ensembles of prediction models optimized on different subsets of the training set. The dataset of 609 X-ray scans was randomly split in 10 subsets of 61 scans each. At the beginning, one of these subsets was selected as testing set and left out of the optimization process. Models were then optimized using different combinations of the nine remaining subsets, and their results were averaged. Models' predictions were averaged at two stages of the pipeline: centerline segmentation maps of several models were averaged to compute single Cobb angles, and Cobb angles predicted by several ensembles were averaged too, acting as an ensemble of ensembles. Results with varying size of ensembles are reported in Table 1. Ensembles with the best results on the left out set of 61 scans were then used for submission on the challenge test set.

3   Experiments

We evaluated the performance of our model by competing in the AASCE challenge (Sect. 2.1). The AASCE challenge uses the symmetric mean absolute percentage error (SMAPE) averaged over the three Cobb angles as main metric. For all three Cobb angles, the SMAPE was computed as:

$$SMAPE = \frac{200}{n} \sum_{i=1}^{n} \frac{\sum_{j=1}^{3} |A_{i,j} - B_{i,j}|}{\sum_{j=1}^{3} A_{i,j} + B_{i,j}}, \tag{2}$$

where n is the number of X-ray scans in the set, $A_{i,j}$ is the manual assessment of the Cobb angle j for the scan i, and $B_{i,j}$ the Cobb angle predicted by the automated method.

To get more insights we also evaluated our results on a subset of the public dataset used for training (Sect. 2.1) with additional metrics. This second test set contained 61 X-ray scans randomly drawn at the beginning of the experiments and not used for optimization. On this set, in addition to the SMAPE, we also computed mean absolute error (MAE) between the manual and automated assessments of Cobb angles. All results are shown in Table 1.

Table 1. Automated predictions of Cobb angles. *Crop* is whether vertebrae at the top and bottom of the image were removed from the image, *Nbr Models* is the number of averaged models, and *Augmentation* is the type of data augmentation on image intensities, where *B & C* are changes of brightness and contrast over the whole image. *MAE 1, 2, 3* are the mean absolute error for the major, upper, and lower Cobb angle in degrees, *MAE Avg* is the MAE averaged over the three angles, *SMAPE* is defined in Eq. (2) and computed over the left out test set of the public dataset (Sect. 2.1), *SMAPE Lb* was computed on the evaluation dataset and reported on the challenge's leaderboard.

Crop	Nbr Models	Augmentation	MAE 1	MAE 2	MAE 3	MAE Avg	SMAPE	SMAPE Lb
No	2	Gaussian	5.28	5.39	6.91	5.86	26.45	27.78
No	4	Gaussian	4.95	5.13	6.55	5.54	24.27	25.79
Yes	4	B & C	3.43	3.91	5.39	4.24	21.03	23.94
Yes	4 & 7	B & C	3.22	3.75	5.23	4.07	20.55	23.67
Yes	4 & 7 & 4 & 4 & 5	B & C	3.18	3.6	5.19	3.99	20.4	22.96

4   Discussion and Conclusion

We proposed an automated method for estimation of Cobb angles from X-ray scans. Contrary to most other approaches, our approach measured Cobb angle directly from the centerline of the spine without requiring the segmentation of individual vertebrae [3,4,8].

Our method has several limitations. In the training dataset all images were initially cropped relatively close to the spine, while in the evaluation dataset the initial regions of interest was substantially larger. This difference caused major issues for the automated spine segmentation, and required manual cropping of the region of interest. Our method also has a long running time due to the smoothing with Euler method. A similar effect may be achieved using Gaussian smoothing. Finally, Cobb angles are not a linear measurement of severity of AIS. The severity actually increases exponentially with the Cobb angle. This is not reflected in our evaluation criteria. Using Cobb angles to assess AIS severity is also in itself limited, as Cobb angles are measured in 2D while AIS is a three-dimensional condition [7]. Using 3D measurements instead could help to assess AIS severity more accurately.

Intrarater or interrater variability have not been measured in our dataset. By inspecting results reported by Horng et al. [4], we computed that, on their set of 32 X-ray scans, the intrarater variability in the manual measurement of Cobb angle by an expert observer had a MAE of 2.0° and interrater variability had a MAE of 3.87° on average. Our results are in-between these intrarater and interrater variabilities, with a MAE of 3.18° and Pearson correlation coefficient of 0.95 between manual and automated measurements of the major Cobb angle in our dataset. Consequently, our method might be suited to replace manual assessment of Cobb angles in clinical practice.

Acknowledgments. This research was funded by the Netherlands Organisation for Health Research and Development (ZonMw) Project 104003005, with additional support of Netherlands Organisation for Scientific Research (NWO) project NWO-TTW Perspectief Programme P15-26.

References

1. Ronneberger, O., Fischer, P., Brox, T.: U-net: convolutional networks for biomedical image segmentation. In: Navab, N., Hornegger, J., Wells, W.M., Frangi, A.F. (eds.) MICCAI 2015. LNCS, vol. 9351, pp. 234–241. Springer, Cham (2015). https://doi.org/10.1007/978-3-319-24574-4_28
2. Weinstein, S.L., Dolan, L.A., Cheng, J.C., Danielsson, A., Morcuende, J.A.: Adolescent idiopathic scoliosis. Lancet **371**(9623), 1527–1537 (2008)
3. Wu, H., Bailey, C., Rasoulinejad, P., Li, S.: Automatic landmark estimation for adolescent idiopathic scoliosis assessment using BoostNet. In: Descoteaux, M., Maier-Hein, L., Franz, A., Jannin, P., Collins, D.L., Duchesne, S. (eds.) MICCAI 2017. LNCS, vol. 10433, pp. 127–135. Springer, Cham (2017). https://doi.org/10.1007/978-3-319-66182-7_15

4. Horng, M.H., Kuok, C.P., Fu, M.J., Lin, C.J., Sun, Y.N.: Cobb angle measurement of spine from X-ray images using convolutional neural network. Comput. Math. Methods Med. **2019**(9), 1–18 (2019)
5. Ioffe, S., Szegedy, C.: Batch normalization: accelerating deep network training by reducing internal covariate shift. In: ICML 2015: Proceedings of the 32nd International Conference on International Conference on Machine Learning, vol. 37, pp. 448–456, July 2015
6. Zeiler, M.D.: ADADELTA: an adaptive learning rate method. arXiv preprint arXiv:1212.5701 (2012)
7. Giannoglou, V., Stylianidis, E.: Review of advances in Cobb angle calculation and image-based modeling techniques for spinal deformities. ISPRS Ann. Photogr. Remote Sens. Spat. Inf. Sci. **III-5**, 129–135 (2016)
8. Mukherjee, J., Kundu, R., Chakrabarti, A.: Variability of Cobb angle measurement from digital X-ray image based on different de-noising techniques. Int. J. Biomed. Eng. Technol. **16**(2), 113–134 (2014)
9. Al Okashi, O., Du, H., Al-Assam, H.: Automatic spine curvature estimation from X-ray images of a mouse model. Comput. Methods Program. Biomed. **140**, 175–184 (2017)

Automated Spinal Curvature Assessment from X-Ray Images Using Landmarks Estimation Network via Rotation Proposals

Rong Tao[1(✉)], Shangliang Xu[1], Haiping Wu[1], Cheng Zhang[2], and Chuanfeng Lv[1]

[1] PingAn Technology (Shenzhen) Co., Ltd., Shanghai 200000, China
{TAORONG037,XUSHANGLIANG223,WUHAIPING855}@pingan.com.cn
[2] Intern, PingAn Technology (Shenzhen) Co., Ltd., Shanghai 200000, China

Abstract. Adolescent idiopathic scoliosis (AIS) is one of the most common type of scoliosis. In current clinical settings, the severity of scoliosis is assessed by evaluating the contralateral blending angle of the spinal cord. Cobb angle is one of the most widely accepted standards for angle measurement. However, the manual measurement of Cobb angle is time consuming and unreliable. In this article, we propose a novel two-stage method that can automatically estimate Cobb angle from vertebrate landmarks. The proposed method uses rotation vertebrate region proposals to increase the accuracy of vertebrate localization in curved spinal region. Our model uses a backbone of ResNet50 combined with FPN for multi-scale region proposal extraction. The rotation proposals are co-registered and fed into stage-two fully convoluted network (FCN) for vertebrate landmarks detection. The performance of proposed method is more robust than traditional landmarks segmentation networks for datasets with large variance, with a SMAPE score of 25.4784.

Keywords: Adolescent idiopathic scoliosis (AIS) · Cobb angle · Vertebrates detection · Landmarks detection

1 Introduction

Adolescent idiopathic scoliosis (AIS) refers to abnormal spinal curvature starts from later childhood or adolescence and continues to adulthood. AIS is one of the most common type of scoliosis, with about 4% occurrence in adolescents [1]. It causes skeletal muscle dysfunction, leading to lower back pain, ventilatory restrictions, or even pulmonary cardiac failure [2]. AIS is preventable, and early diagnosis and proper intervention in adolescents are critical for controlling abnormal spinal curve progression.

Patients with AIS may experience mild or no pain in the early stage. Clinically, the diagnosis of AIS are relying on curvature assessment on lateral spinal X-ray images, and one of the widely used standard is Cobb angle. However, the manual measurement of Cobb angles is time consuming and labor intensive. It can be challenge for experienced clinicians due to anatomical variations and low contrast of X-ray images.

R. Tao, S. Xu and H. Wu—Joint main authors and equally contribute to the project.

© Springer Nature Switzerland AG 2020
Y. Cai et al. (Eds.): CSI 2019, LNCS 11963, pp. 95–100, 2020.
https://doi.org/10.1007/978-3-030-39752-4_11

Recent advances in deep learning show the advantages of automatic spinal curvature assessment. Wu *et al.* [3, 4], Sun *et al.* [5] and Galbusera *et al.* [6] proposed one-stage bottom-up approaches to extract vertebrate landmarks or Cobb angle directly from spine images. One-stage approaches can incorporate structural information of whole spinal region but are easily affected by intra domain variance. One the other side, Horng *et al.* [7] used two-stage approach to first locate the rectangular vertebrate regions and then estimate vertebrate landmarks in each region. Two-stage approaches split the task of landmarks detection on the whole image into two easier sub-tasks: vertebrate region localization and landmarks detection on each region proposal. On problem with existing two-stage methods is that rectangular vertebrate proposals are not suitable for curved spinal cords. Since vertebrates are not horizontal when scoliosis occurs, and accuracy of both two sub-tasks can be affected due to region misalignment.

To address this problem, we propose a two-stage automated spinal landmarks detection network based on rotational regional proposals of vertebrates. In stage one, the proposed network detects location of vertebrates using rotated rectangular regions. In stage two, each vertebrate region undergoes rotation co-registration using rotation angle from previous stage. Landmarks detection are then performed on aligned proposals. Finally, we estimate Cobb angle using detected vertebrate landmarks.

2 Proposed Method

2.1 Vertebrates Detection via Rotational Proposals

We adopt feature pyramid network (FPN) as network's backbone (see Fig. 1), which performs multiscale feature extraction and gives regional proposals on each scale. In the training stage, the input ground truth of each rectangular vertebrate bounding box contains 5 parameters (x, y, w, h, θ). Among them, (x, y) and (w, h) describe bounding box center location and dimension, respectively. Rotation angle θ, ranging from $-\frac{\pi}{2}$ to $\frac{\pi}{2}$, is about the angle of tilted bounding box with respect to x-axis with rotation center fixed at (x, y). In addition, we design rotational anchors at 5 scales, 3 ratios and 3 rotational angles. We use minimum mean square loss for rotation angle regression. The rest parameters of bounding box are estimated as in [8]

2.2 Rotational ROI Align and Landmarks Detection

We propose to use rotational region of interest (ROI) align on vertebrate proposals (see Fig. 1). Rotated vertebrate proposals are adjusted using rotation angles regressed from previous stage. These proposals then undergo ROI align into fixed size feature maps as stated in [8]. Finally, they are sent to fully convolutional network (FCN) for landmarks detection.

3 Experiments and Results

3.1 Implementing Details

Dataset. The proposed study is based on AASCE2019 whole spine X-ray dataset, including 481 train images, 128 validation images, and 98 test images. Images in train

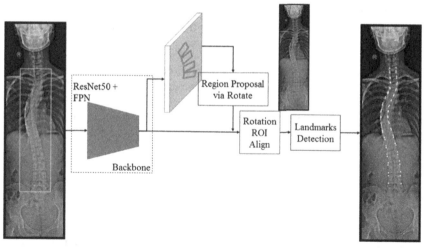

Fig. 1. Pipeline of proposed method. We use ResNet50 combined with FPN as backbone for multiscale feature extraction. Rotational region proposals are extracted from stage one and fed into FCN for landmarks detection.

and validation set consist of 68 manually annotated landmarks (4 landmarks for each of the 17 thoracic vertebrates). Trainset and validation set are manually cropped to remove head and pelvis floor, whereas test set images have increased field of view from upper femur to head. In addition, image contrast and texture of test set are visually different from train and validation set. The intra domain variance among three datasets, particularly the presence of head and pelvis floor in test set, has a significant impact on model performance (see Fig. 2(a) and (b)). To solve this, we propose a standardized image preprocessing routine to reduce the influence of dataset variance.

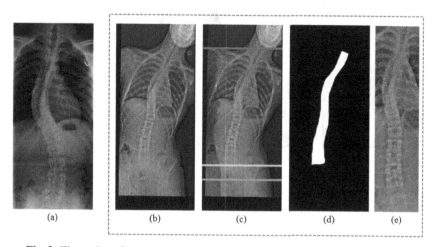

Fig. 2 Illustration of preprocessing steps for spine region isolation and refinement.

Spine Region Isolation. We first isolate the area of spine region using horizontal intensity projection to remove head and pelvis floor (see Fig. 2(c)). The intensity thresholds for different body parts are determined by statitcal evaluation of whole dataset.

Spine Region Refinement with Cord Segmentation. After obtaining spine region, we perform spinal cord segmentation and midline extraction (see Fig. 2(d)), which are used for spinal cord co-registration and false-positive key points suppression. Finally, spine region is further refined with locked aspect ratio 3.5 between image height and width (see Fig. 2(e)).

3.2   Results

Figure 3. shows experiment results of vertebrate localization and landmarks detection of test dataset. Our model predicts the location of each vertebrate using rotation bounding box (as in Fig. 3(b)). Despite significant difference among train, validation and test dataset, our model achieves a recall of >98% in vertebrate detection and a precision of >99% after post-processing using spinal cord mask. The proposed vertebral region and co-registered using rotation angle and used for landmarks detection. Each region proposal yields 4 landmarks, which are projected back into original image for Cobb angle estimation.

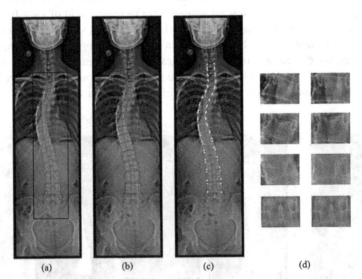

Fig. 3. Qualitative experiment results of test dataset. (a) shows the original image with refined spine region. (b) shows the rotational bounding box results from stage one region proposal network. (c) shows vertebrate landmarks segmentation results from stage two landmarks detection module. (d) illustrates vertebrate patches before and after rotation angle adjustment.

We compare our method with V-Net and Cascade Pyramid Network (CPN). V-Net is a robust architecture that succeed in many medical image applications. For V-Net, we divide the landmarks into four groups (upper left, lower left, upper right, and lower left), according to the location of landmarks to corresponding vertebrate. CPN is a powerful two-stage landmark detection network that demonstrate its ability in human landmarks detection. We use similar settings for CPN with our method. Experiments results shows that our method achieves better performance that the other two, with a score of 25.4784 for proposed method, 30.7135 for V-Net and 40.7873 for CPN ((as in Table 1). By visual inspection, we find that rotation region proposals fit vertebrate region better and non-rotate proposals, thus reduce noises from non-spinal regions and therefore increase the accuracy of landmark detection.

Table 1. Quantitative comparison of three different methods.

Model	V-Net	CPN	Our method
SMAPE	30.7135	40.7873	25.4784

4 Conclusion

In this paper, we introduce an end-to-end deep convolutional neural network for spinal landmarks detection and Cobb angle estimation. Our method is motivated by MaskR-cnn but customized for the task of rotated vertebrate detection using rotational region proposals. Compared to other methods, the proposed method is more robust to intra domain variance. Rotation region proposals provide a better fitting of vertebrates in curved spinal regions, which are of greater clinical importance than straight spinal area. Given the accurate detection of vertebral regions, our model has a lower false-positive rate of estimated landmarks.

The task of Cobb angle estimation intrinsically requires locating the first and last thoracic vertebrate. However, our model is more likely to detect all existing vertebrates on image. We have incorporated preprocessing steps such as head and neck removal, which can partly solve this problem. In practice, however, the last cervical vertebrate is often mis-recognized as the first thoracic vertebrate. One future direction is to introduce additional network for more accurate localization of the first and last thoracic vertebrate.

Overall, we have proposed a novel and robust framework for spinal landmarks detection. The proposed method demonstrates its ability in dataset with large variance and can potentially be used for other medical image applications.

References

1. Konieczny, M.R., Senyurt, H.: Epidemiology of adolescent idiopathic scoliosis. J. Child. Orthop. **7**(1), 3–9 (2013)
2. Martinez-Llorens, J., Ramirez, M., Colomina, M.J.: Muscle dysfunction and exercise limitation in adolescent idiopathic scoliosis. Eur. Respir. J. **36**(2), 393–400 (2010)

3. Wu, H., Bailey, C., Rasoulinejad, P., Li, S.: Automated comprehensive adolescent idiopathic scoliosis assessment using MVC-Net. Med. Image Anal. **48**, 1–11 (2018)
4. Wu, H., Bailey, C., Rasoulinejad, P., Li, S.: Automatic landmark estimation for adolescent idiopathic scoliosis assessment using boostnet. In: Descoteaux, M., Maier-Hein, L., Franz, A., Jannin, P., Collins, D.L., Duchesne, S. (eds.) MICCAI 2017. LNCS, vol. 10433, pp. 127–135. Springer, Cham (2017). https://doi.org/10.1007/978-3-319-66182-7_15
5. Sun, H., Zhen, X., Bailey, C., Rasoulinejad, P., Yin, Y., Li, S.: Direct estimation of spinal Cobb angles by structured multi-output regression. In: Niethammer, M., et al. (eds.) IPMI 2017. LNCS, vol. 10265, pp. 529–540. Springer, Cham (2017). https://doi.org/10.1007/978-3-319-59050-9_42
6. Galbusera, F., Bassani, T., Costa, F.: Artificial neural networks for the recognition of vertebral landmarks in the lumbar spine (2017)
7. Horng, M.H., Kuok, C.P., Fu, M.J.: Cobb angle measurement of spine from x-ray images using convolutional neural network. Comput. Math. Methods Med. (2019)
8. Kaiming, H., Georgia, G., Piotr, D.: Mask R-CNN. IEEE Trans. Pattern Anal. Mach. Intel. 1 (2018)
9. Chen, Y., Wang, Z., Peng, Y., et al.: Cascaded Pyramid Network for Multi-Person Pose Estimation (2017)

A Coarse-to-Fine Deep Heatmap Regression Method for Adolescent Idiopathic Scoliosis Assessment

Zhusi Zhong[1], Jie Li[1], Zhenxi Zhang[1], Zhicheng Jiao[2], and Xinbo Gao[1(✉)]

[1] School of Electronic Engineering, Xidian University, Xi'an 710071, China
xbgao@mail.xidian.edu.cn
[2] Perelman School of Medicine, University of Pennsylvania, Hamilton, PA 19104, USA

Abstract. Spinal anterior-posterior x-ray CT imaging is an appealing tool to aid diagnosis and elucidate Adolescent Idiopathic Scoliosis (AIS) Assessment. In this paper, we propose an automatic detection method for AIS assessment from X-ray CT images. Our deep learning based coarse-to-fine heatmaps regression method achieves symmetric mean absolute percentage error (SMAPE) of 24.7987 in the grand challenge AASCE 2019.

1 Introduction

Adolescent idiopathic scoliosis (AIS) is the most common form of scoliosis and typically affects teens. It is a medical condition in which a person's spine has a sideways curve. The curve of the spine presents "S"- or "C"-shape over three dimensions. The conditions of some patients are not stable, and the degree increases over time. Mild scoliosis does not typically cause problems, but severe cases can interfere with breathing. The assessment has significant impacts on statistics and prediction of the spinal condition, which can help to decide the early treatment plan and prevent the deterioration of the condition.

The AIS assessment is a standard tool to quantitatively analyze the condition. The evaluation is based on some landmarks around the vertebras in 2D vertical section X-ray image of human trunk. The Cobb angle is a standard measurement of bending disorders of the vertebral column. The angles are calculated on the annotated landmarks of the posteroanterior (back to front) X-ray images. The evaluation selects the most tilted vertebra at the top and bottom of the spine, and calculates the intersection angles of the selected vertebras. But in clinical, the landmarks are annotated manually, which remains a time-consuming work for an experienced doctor. Because of the anatomical differences across organizations, the manual annotation in the X-ray images is extremely subjective to observer variability. The accuracy of assessment usually has a great influence on the treatments. Therefore, an automatic annotation and calculation method would release doctors from the time-consuming work and especially avoid the observation errors. Li et al. [1–4] hold the Accurate Automated Spinal Curvature Estimation as a grand challenge in MICCAI 2019.

The AIS assessment is directly based on spinal landmarks. But in the evaluation, the Cobb angle is designed to selecting the most tilted vertebra, and the endplates are

© Springer Nature Switzerland AG 2020
Y. Cai et al. (Eds.): CSI 2019, LNCS 11963, pp. 101–106, 2020.
https://doi.org/10.1007/978-3-030-39752-4_12

generally parallel for each vertebra, so the Cobb angle is not sensitive to small local deviation in landmark coordinates. In this paper, we propose a novel deep learning framework for automatically AIS assessment. Our proposed method is improved from the landmark heatmap regression method. In our framework, the deep learning models regress heatmaps from coarse to fine in 2 stages, informing global configuration as well as accurately describing the local appearance, similar to what we did in [5].

2   Method

Overall Framework: As shown in Fig. 1, the overall framework for landmark detection includes 2 stages. The global stage takes the peeling cropped image as input, the U-net [6] A regresses the whole spine mask. The image multiplied with the mask and inputs to Mask r-cnn [7], whose outputs contain box masks, bounding boxes and tag labels. In the local stage, we extend the bounding box and crop the region image contained the box. The local region image inputs to U-net D and U-net E separately, the outputs are multiplied together. The 4 highlights in the 4 channels heatmaps are obtained as the landmarks of one vertebral column. The local stage procedure traverses the bounding boxes and obtains the spine landmarks. Based on the previous work, we proposed a different landmark obtaining method. To increase the robustness, we modify the landmarks with polynomial curve fitting, to adjust the unusual landmarks and decrease the outliers.

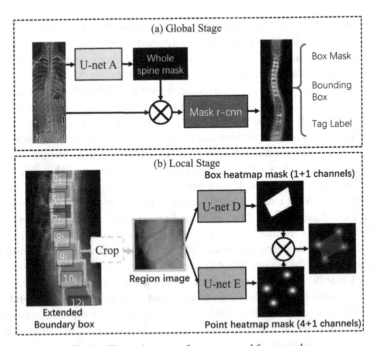

Fig. 1. The main parts of our proposed framework.

Global Stage Masks Regression: We train 3 U-net models to regress 3 kinds of heatmaps, as shown in Fig. 2. The 3 models take whole spine images as input. The U-net A is trained to regress Whole Spine Mask (WSM). The U-net B is trained to regress the Target Spine Mask (TSM). The outputs of these 2 models are highlight regions of the spine, but their edges are with different blur extension. The first channel is the target channel as shown in Fig. 2, and the last channel is the background which is all 1 matrix subducts the sum of the target channels. The WSM and TSM indicate the location of the spine, they can filter out the false positive regions while inferring. The U-net C is trained to regress the Box Masks. The box masks are the center region in the 4 landmarks of each vertebral column, and the first 17 channels contain 17 heatmaps separately for the vertebras in the given data. The last channel is the shared background, in order to handle the class-imbalance problem while training. The U-net C separates the connective vertebras into channels, the highlights in channels can easily represent their locations. The 3 models are trained separately and are embedded in the framework described in the following section.

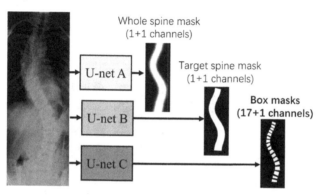

Fig. 2. 3 U-net models regress 3 different kinds of heatmaps.

Peeling Crop: The released test data is much more irregular and wilder than the training data. We propose a data regularization method for data preprocessing, in order to fit the trained model while inferring. The proposed data preprocess method is a progressive procedure, gradually narrowing the unwanted margins, like peeling onion piece by piece to get the wanted inner part. We use the 3 trained U-net models in the peeling crop procedure. The trained U-net A model takes the wild test data as input, to regress the WSM. Although there is a huge deviation between the train data and the input data, the model can still locate the coarse spine region but with marginal false detection. The U-net B and U-net C take the test data multiplied the first channel of the WSM, to regress TSM and box masks. To obtain the main target spine region, the first channel of TSM multiplies with the first 17 channels of box masks. Each channel obtains the highlight location. We apply the outlier detection and the continuity judgement to the 17 coordinates, to figure out the superior endplate and the inferior endplate. The detected region box is based on the 2 endplates' location and with the fixed height-width ratio, which is the red dotted box as shown in Fig. 3. We extend the detected box with the remained margin, in case of the over cropping in the target spine. The cyan dotted box in the Fig. 3 is the extended region. The next iteration uses the cropped image in the extended region. The peeling crop iterates the cropping and the predicting, until the detected region box is narrowing slowly. We compare the narrowing distance of 2 iterations and set the maximum iteration number, to stop the iteration procedure. As shown in Fig. 4, when the region is getting closer to the training data, the box masks are getting more precise and separable in channels. The peeling crop processes the testing data once, before the inferring in the main framework.

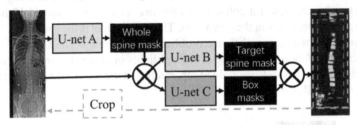

Fig. 3. The peeling crop is the data prepressing for the test data. (Color figure online)

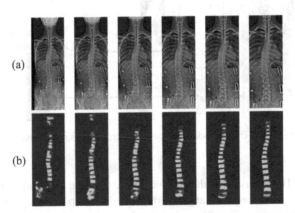

Fig. 4. The cropped image in the extended spine region as shown in row (a) and their box masks predictions in row (b). They are progressively adjusting in the peeling crop iteration.

Global Stage Detection: The vertebras are similar locally, so we model the vertebra detection problem as an instance segmentation problem. As shown in Fig. 1a, our global stage detection is based on the Mask r-cnn, which has shown well-performance in the instance segmentation subject. We transfer the coordinates to box masks and bounding boxes. The tag labels are the up-down order index of vertebras in the annotated spine. The U-net A takes the peeling cropped image as input, and then the output WSM multiplies with the image. The Mask r-cnn is trained to predict the box masks, bounding boxes and tag labels. The model extracts global information for the local patch detection.

Local Stage Heatmap Regression: We got the coarse bounding boxes, which indicate the boundaries of vertebras. In local stage, we want to obtain the landmarks for each vertebra. We use multi channels heatmaps regression method similarly to what we did in [5]. As shown in Fig. 1b, the local stage contains U-net D and U-net E. These 2 models are trained with the region patch images, to regress the box heatmaps and the point heatmaps. The box heatmaps are 2 channels, the first channel is the box mask of the vertebra located at the center of the extended bounding box, the second channel is the background. The point heatmaps are 5 channels, the 4 landmarks of one vertebra are modeled as 4 channel heatmaps. A 2D Gaussian distribution locates at the center of the landmark in the first 4 channels, and the last channel is the sheared background. While inferring, the points heatmaps are masked separately by the first channel of the box heatmaps. The overlapping region of the first 4 channels contains the locations of the landmarks. Each coordinate is obtained as the mean position of the pixels whose values are greater than 0.5. The 4 coordinates of 1 vertebra are obtained in the first 4 overlapping channels. The local stage procedure traverses the bounding boxes and obtains the whole spine landmarks.

Polynomial Curve Refine: We find that the Cobb angle is sensitive to the unusual landmarks, and these error predictions are usually protrusive and uneven in their vertebral neighbors. But the midline of the prediction shows the reliable curve trend of the spine. As the post-processing, we want to use the curve trend to fine-tune these error predictions. There are 2 sets of coordinates here used to fit the spine curve, which are the centers of the bounding boxes from Mask r-cnn and the mean positions of the box masks from U-net C. We use the least square method to fit the 9th degree polynomial coefficient, and the curve segments of the 2 endplates are 1st degree polynomial coefficient. We iterate the fitting procedure, while filtering out those landmarks far from the curve. The final coordinates are refined by the curve. The 17 center labels of bounding boxes which matching to the box masks from U-net C, are selected as the target vertebras. The horizonal connections of their coordinates intersect the curve. The vertical line in the intersections should go thought the 2 coordinates in an ideal situation. The unusual coordinates rotate on their intersections closing to the vertical lines.

3 Results

We report the results of two stages obtained by the proposed framework. The data is provided by the grand challenge AASCE 2019 [8], and the results are the online evaluations on the unlabeled test dataset. The evaluation is the score based on symmetric mean absolute percentage error (SMAPE) (Table 1).

Table 1. The stage-wise results of the proposed framework.

Stage	SMAPE
Local stage	26.4455
Curve refine	24.7987

4 Discussion

In conclusion, we propose a coarse-to-fine heatmap regression method for Adolescent Idiopathic Scoliosis Assessment. As data preprocessing, the Peeling Crop strategy is different from the direct region prediction methods, it adjusts the target regions progressively. The global and local parts of our framework are trained separately, it reduces the GPU memory cost and they are motivated in 2 different strategies. The 2 stages inform global configuration as well as accurately describing local appearance, then the local stage regresses the point heatmaps to obtain the landmark locations. We propose a Polynomial Refine method as our post-processing to refine the local stage coordinates. The results show that our method is effective in AIS Assessment.

References

1. Wu, H., Bailey, C., Rasoulinejad, P., Li, S.: Automatic landmark estimation for adolescent idiopathic scoliosis assessment using BoostNet. In: Descoteaux, M., Maier-Hein, L., Franz, A., Jannin, P., Collins, D.L., Duchesne, S. (eds.) MICCAI 2017. LNCS, vol. 10433, pp. 127–135. Springer, Cham (2017). https://doi.org/10.1007/978-3-319-66182-7_15
2. Wang, L., Qiuhao, X., Leung, S., Chung, J., Chen, B., Li, S.: Accurate automated Cobb angles estimation using multi-view extrapolation net. Med. Image Anal. **58**, 101542 (2019)
3. Chen, B., Xu, Q., Wang, L., Leung, S., Chung, J., Li, S.: An automated and accurate spine curve analysis system. IEEE Access **7**, 124596–124605 (2019)
4. Wu, H., Bailey, C., Rasoulinejad, P., Li, S.: Automated comprehensive Adolescent Idiopathic Scoliosis assessment using MVC-Net. Med. Image Anal. **1**(48), 1–11 (2018)
5. Zhong, Z., Li, J., Zhang, Z., Jiao, Z., Gao, X.: An attention-guided deep regression model for landmark detection in cephalograms. arXiv preprint arXiv:1906.07549 (2019)
6. Ronneberger, O., Fischer, P., Brox, T.: U-Net: convolutional networks for biomedical image segmentation. In: Navab, N., Hornegger, J., Wells, W.M., Frangi, A.F. (eds.) MICCAI 2015. LNCS, vol. 9351, pp. 234–241. Springer, Cham (2015). https://doi.org/10.1007/978-3-319-24574-4_28
7. He, K., Gkioxari, G., Dollár, P., Girshick, R.: Mask R-CNN. In: Proceedings of the IEEE International Conference on Computer Vision, pp. 2961–2969 (2017)
8. https://aasce19.grand-challenge.org

Spinal Curve Guide Network (SCG-Net) for Accurate Automated Spinal Curvature Estimation

Shuxin Wang, Shaohui Huang, and Liansheng Wang[⊠]

Department of Computer Science, School of Informatics,
Xiamen University, Xiamen 361005, China
sxwang@stu.xmu.edu.cn, {hsh,lswang}@xmu.edu.cn

Abstract. Cobb angle plays an important role in the diagnosis of scoliosis, which can effectively quantify the degree of scoliosis. Manual measurement of Cobb angles is time-consuming, and the results are also heavily affected by the expert's choice. In this paper, we propose a spine curve guide framework to directly regress the cobb angle from single AP view X-rays images. We firstly design a segmentation network to accurately segment two spine boundary, and then aggregate the obtained boundary scoremap with the original spinal X-rays images to input another angle estimation network to make high-precision regression prediction for cobb angle. We evaluate our method in the AASCE19 challenge, and our result achieves 22.1775 SMAPE that shows strong competitiveness compared to other excellent methods.

1 Introduction

Scoliosis is a disease caused by abnormal curvature of the spine, the curve of the scoliosis usually appears as an "S" or "C" shape in a posteroanterior (back to front) X-ray. In the diagnosis and treatment decisions of scoliosis, cobb angle is widely used to evaluate the degree of curvature of the spine [4], which is shown in Fig. 1(a).

Traditional methods of measuring Cobb angles employed hand-crafted landmarks of spine to measure the cobb angle which is time-consuming. And these manual landmarks are affected by many factors such as the selection of vertebrae, the bias of observer, as well as image quality [1]. To improve the accuracy of cobb angle and eliminate manual impact, recent works are more focus on the end-to-end automated estimation methods based on deep convolutional neural networks. Wu *et al.* [6] integrate the robust ConvNet features with statistical information to adapt to the variability in X-ray images. After that, Wu *et al.* [5,7] aggregate multi-view information from both AP and LAT x-rays, aiming at utilizing the structural dependencies of the two views. However, these works are directly regress the cobb angle from the raw X-rays, which don't make good use of clinical priori knowledge. More recently, Chen *et al.* [1] design a landmark-net

© Springer Nature Switzerland AG 2020
Y. Cai et al. (Eds.): CSI 2019, LNCS 11963, pp. 107–112, 2020.
https://doi.org/10.1007/978-3-030-39752-4_13

to learn the spinal boundary features from landmarks for comprehensive scoliosis assessment.

However, Chen *et al.* [1] only use landmarks for indirect Cobb angle calculation, which makes few contributions to the learning process of regression network. In this paper, we propose a spine curve guide framework (SCG-Net) to more accurately predict the cobb angle by utilizing the spine curve segmentation result as correction information to help network focus on more useful features during later regression task. Consider that the spine curve is capable of reflecting the degree of scoliosis intuitively, we firstly design a segmentation network to learn boundary of spine. Unlike latest work [1] that obtains auxiliary information from sporadic spinal landmarks (see Fig. 1(b)), we found that continuous spine boundary curve (see Fig. 1(c)) carries more complete analysis for the spine. In later regression task, the network directly predicts the cobb angles based on original input images and the spine curve map generated by segmentation model and the spine curve map could help the regression network focus on more useful features. To verify the effectiveness of the presented SCG-Net, we evaluate it on the test set of AASCE2019, and the leaderboard demonstrates our method achieves excellent performance.

2   Method

Figure 2 illustrates the pipeline of our method. Firstly, we learn the spine boundary curve from given X-rays images by a boundary segmentation network. Then the angle regression network takes X-ray images and obtained spine curve maps as inputs to produce the final prediction for cobb angle.

2.1   Boundary Segmentation Network

Based on clinical knowledge, we know that cobb angle reflects the crook extent of spine, therefore we design a segmentation network to obtain the spinal boundary. The segmentation network is designed as Unet [3] structure that consists of an encoder and a decoder and we add two skip-layer connections at 1X and 4X levels to refine the segmentation results by aggregating the low-level features and high-level semantic features. In order to learn more complete spine edge, we apply the Pyramid Pooling Block (PP block) [8] at the bottom of network to generate global information at different scales.

2.2   Angle Regression Network

Unlike classical classification and segmentation tasks, directly regress an impalpable cobb angle value from the raw X-ray image is lack of interpretability, even for experts. In fact, the cobb angle is usually calculated from spine curve, (see Fig. 1(a)(c)), therefore, we utilize the spinal boundary segmentation to build a relation between raw X-ray images and the feature learning of cobb angle to boost the network's capability of understanding cobb angle.

AP X-ray Image

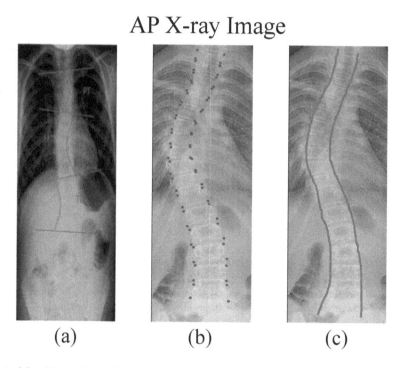

(a)   (b)   (c)

Fig. 1. (a) cobb angles in the posteroanterior (back to front) X-ray (b) given landmarks of spine, (c) boundary curve of spine

Our angle regression network takes the raw X-ray images and their corresponding segmentation results as inputs and directly produces three predictions for cobb angles. The main body of our regression network is DenseNet [2], which connects each layer to every other layer in a feed-forward fashion and shows excellent performance in a lot of tasks.

3   Experiments

3.1   Dataset and Data Augmentation

The AASCE2019 challenge supplies 609 spinal anterior-posterior X-ray images as train set and each image contains three cobb angle tags (PT, MT, TL). In addition, organizers provide corresponding landmark information for each image in train phase (see Fig. 1(b)). In test phase, all participates need evaluate their method on 98 images without landmark information.

For the train dataset, we make little data augmentation except for random crop and resize, because we are concerned that excessive transformations may affect the accuracy of the angle tag. For segmentation task, we firstly connect given landmark points together to generate two continuous spine boundaries and then we make dilation operation on the curve to generate final segmentation mask.

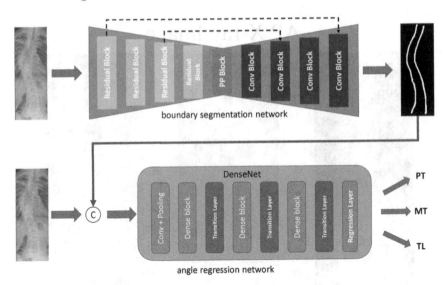

Fig. 2. The workflow of our method, which consists of two networks. The above network generates segmentation maps for spine boundary curves and the one below takes the original inputs as well as obtained spine curves mathe crook extent of spineps as input to make the final prediction for cobb angles.

3.2   Evaluation Metrics

In this challenge, the symmetric mean absolute percentage error (SMAPE) is used to evaluate the effectiveness of each method, which is defined as following:

$$SMAPE = \frac{1}{N} \sum_{N} \frac{SUM(|A - B|)}{SUM(|A + B|)} \times 100\% \qquad (1)$$

where A is ground truth, B is predicted result, N is the number of test data.

3.3   Challenge Result

The evaluation result of the on-site challenge of AASCE 2019 is listed in Table 1, and our method demonstrates good performance and strong robustness. Our results finally rank 5th place on the public leaderboard and there is only a slight gap between our results and the first place.

Table 1. There are 257 teams participated in this challenge and 78 teams submitted their results. Our method ranks the 5th place.

Team	SMAPE	Rank
mr.waste(TencentX)	21.7135	1
Walker(iFLYTEK)	22.1658	2
AIchallenge2019	22.1729	3
Haru1994	22.1746	4
Our method	22.1775	5
......	......	...
soeci92	50.6884	78

4   Conclusion

In this paper, we present a novel cobb angle estimation framework (SCG-Net) that introduces the spine curve map as a guidance to help angle estimation network understand this task better. Finally, our method ranks 5th in the AASCE2019 challenge and there is only a slight gap between our results and the first place.

Acknowledgement. This work was supported by National Natural Science Foundation of China (Grant No. 61671399) and by the Fundamental Research Funds for the Central Universities (Grant No. 20720190012).

References

1. Chen, B., Xu, Q., Wang, L., Leung, S., Chung, J., Li, S.: An automated and accurate spine curve analysis system. IEEE Access **7**, 124596–124605 (2019)
2. Huang, G., Liu, Z., Van Der Maaten, L., Weinberger, K.Q.: Densely connected convolutional networks. In: Proceedings of the IEEE Conference on Computer Vision and Pattern Recognition, pp. 4700–4708 (2017)
3. Ronneberger, O., Fischer, P., Brox, T.: U-Net: convolutional networks for biomedical image segmentation. In: Navab, N., Hornegger, J., Wells, W.M., Frangi, A.F. (eds.) MICCAI 2015. LNCS, vol. 9351, pp. 234–241. Springer, Cham (2015). https://doi.org/10.1007/978-3-319-24574-4_28
4. Vrtovec, T., Pernuš, F., Likar, B.: A review of methods for quantitative evaluation of spinal curvature. Eur. Spine J. **18**(5), 593–607 (2009)
5. Wang, L., Xu, Q., Leung, S., Chung, J., Chen, B., Li, S.: Accurate automated cobb angles estimation using multi-view extrapolation net. Med. Image Anal. **58**, 101542 (2019)
6. Wu, Hongbo, Bailey, Chris, Rasoulinejad, Parham, Li, Shuo: Automatic Landmark Estimation for Adolescent Idiopathic Scoliosis Assessment Using BoostNet. In: Descoteaux, Maxime, Maier-Hein, Lena, Franz, Alfred, Jannin, Pierre, Collins, D.Louis, Duchesne, Simon (eds.) MICCAI 2017. LNCS, vol. 10433, pp. 127–135. Springer, Cham (2017). https://doi.org/10.1007/978-3-319-66182-7_15

7. Wu, H., Bailey, C., Rasoulinejad, P., Li, S.: Automated comprehensive adolescent idiopathic scoliosis assessment using MVC-Net. Med. Image Anal. **48**, 1–11 (2018)
8. Zhao, H., Shi, J., Qi, X., Wang, X., Jia, J.: Pyramid scene parsing network. In: Proceedings of the IEEE Conference on Computer Vision and Pattern Recognition, pp. 2881–2890 (2017)

A Multi-task Learning Method for Direct Estimation of Spinal Curvature

Jiacheng Wang[1], Liansheng Wang[1], and Changhua Liu[2(✉)]

[1] Department of Computer Science, Xiamen University, Xiamen, China
[2] Department of Medical Imaging,
The Chenggong Hospital Affiliated to Xiamen University, Xiamen 361005, China
liuxingc@126.com

Abstract. In scoliosis diagnosis and treatment, estimation of spinal curvature plays an important role. Compared with the traditional method, which is time-consuming and unreliable, automated estimation has been more and more popular. But it remains to be such a great challenge that direct estimation has poor precision due to the lack of information. To meet this challenge, we propose a Multi-Task learning method with pyramidal feature aggregation. Our method is one-stage. It means that we can directly estimate the angles without detecting landmarks. To enhance the feature extraction and collect more information, we make the fusion of the pyramidal features and extend the base model by adding an extra branch for spinal segmentation. We evaluate our method on the validation set from the challenge (Accurate Automated Spinal Curvature Estimation, MICCAI 2019) and obtain a symmetric mean absolute percentage error of 12.97.

Keywords: Spinal curvature · Pyramidal features · Spinal segmentation

1 Introduction

Cobb angle is a measurement of the bending deformity of the vertebral column. It has been widely used for scoliosis treatment, including structural, lateral, rotated curvature of the spine. Large reports show that there is a sustained increment of the prevalence rate of scoliosis. Therefore, it's likely essential to have a reliable estimation of Cobb angles.

In clinical practice, each X-ray contains seventeen vertebras. For each one of them, doctors find four landmarks and draw the outline accordingly. Then they locate the key vertebras and calculate the angles. The main obstacle is that this method is highly time-consuming and can be easily mistaken.

To address this problem, researchers have designed lots of automated frameworks. Two-stage methods like [9] simulate clinical practice. They firstly segment anatomical structures and then estimate the measurement based on segmentation. Some recent studies adopt the detector, instead of the segmentation model,

© Springer Nature Switzerland AG 2020
Y. Cai et al. (Eds.): CSI 2019, LNCS 11963, pp. 113–118, 2020.
https://doi.org/10.1007/978-3-030-39752-4_14

to observe the key structure [5]. But these studies are limited by the selection of key vertebras and bias of different operators. Direct estimation methods [6–8] aim to obtain the functional connection of medical images and clinical measurement. While these studies are limited by the multi-type views (anterior-posterior and lateral X-ray images). In the case of a single view, they perform much worse due to the insufficient use of global information.

For the purpose of full use of global information, we propose a direct estimation method based on pyramidal feature aggregation. In contrast to earlier direct methods, we add an extra branch to predict spinal masks inspired by [1], which can benefit the angles' estimation. To further enhance the feature extraction, we make a fusion of the decoder's feature map with multiple scales. Overall, our method achieves the highest symmetric mean absolute percentage error (SMAPE) of 12.8 on the 125 scans from the challenge.

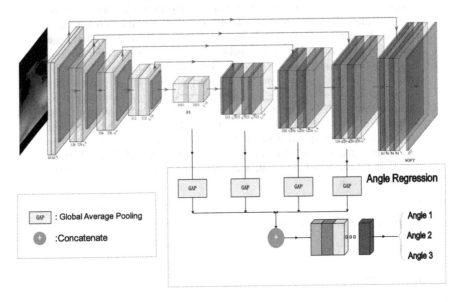

Fig. 1. Overall of our framework. The network learns multi-scale features from single view (AP) via pyramidal feature aggregation and multi-task learning.

2   Methods

2.1   System Framework

As is shown in Fig. 1, the network is based on the encoder-decoder structure [3]. The last feature map of the encoder is transformed into a vector via global average pooling (GAP), and the biggest feature map of the decoder outputs the mask via activation function (Softmax). In an attempt to make full use of a decoder, we add GAP into each stage of the decoder except the last one.

For the different lengths, the feature vectors are simply concatenated and output the three angles through the dense layer.

For network training, our loss function includes mask loss and angle loss which can be defined as:

$$loss = loss_{mask} + loss_{angle} \qquad (1)$$

Dice similarity coefficient (DSC) is a metric function to measure the degree of similarity and always used in the medical image segmentation, so we calculate the mask loss by DSC. As for the Cobb angle, we choose SMAPE, which is usually used to evaluate the angles' estimation.

$$loss_{mask} = 1 - \frac{2|X \cap Y|}{|X| + |Y|} \qquad (2)$$

$$loss_{angle} = 1 - \frac{1}{N} \sum_{N} \frac{SUM\left[|A - B|\right]}{SUM\left[A + B\right]} \times 100\% \qquad (3)$$

Here A and X are ground truth, B and Y are predicted result, N is the number of test data.

2.2 Image Pre-processing

The direct estimation method hasn't taken it into consideration that different ratio (rate of height and width) will obtain different Cobb angles (see Fig. 2). There is a certain problem while resizing all the images into the same size. For instance, if the original ratio is smaller than the model's input ratio, Cobb angles will be smaller in the result. Hence we keep the ratio while resizing and use padding to fit the input shape. What's more, we augment the training data by randomly shifting resized images.

3   Experiments

3.1 Datasets and Implementation Details

Datasets are collected from the challenge (Accurate Automated Spinal Curvature Estimation, MICCAI 2019), and it's composed of three parts, among which only train and validation sets' annotations are provided. As a result, we take the experiments only on the two sets. We count the three angles' value and draw the histogram as Fig. 3. From top to bottom, we number the Cobb angles. It's apparent that the angle above is more likely to be bigger than the angle below. From the graph, we can also see that most angles are small.

Although samples in the training set may come from the same patient, considered that there is still bias due to interval of operation, we randomly select 10% samples from the public train set as the validation and take public validation set as our test set. For all experiments, we use the same optimizer ($Adam$) and the learning rate ($1e-4$). By initialized with 100 epochs, the training process is terminated by a strategy called $Earlystop$.

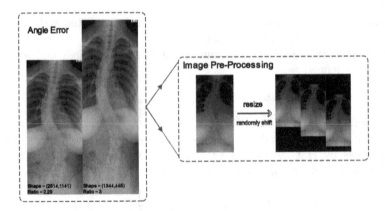

Fig. 2. Purpose of the pre-processing method. There is some error while changing original image into different ratio, which may lead to lose of information.

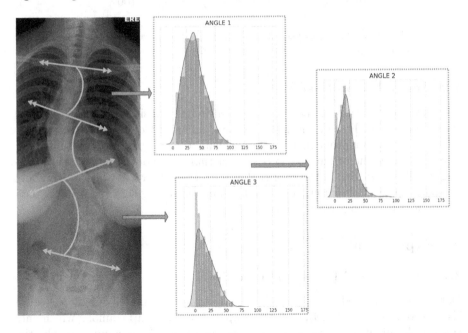

Fig. 3. Count of the three angles' value. From top to bottom we number the angles as ANGLE 1–3. Blue box represents the histogram and red line is the univariate or bivariate kernel density estimation. The x-axis means angle's value and we set the hist bins as 25. (Colr figure online)

3.2   Results

Through the same setting described above, we have taken a series of experiments to make the discussion. First of all, we talk about the impact of different encoder: Vgg11 [4], ResNet-50 [2]. Thanks to the single channel of X-ray images, weights pre-trained on ImageNet cannot be transferred into this task. Thus we train both networks from scratch and haven't adopted deeper encoder. As the second and third rows show, there is a significant difference between the two conditions. Vgg11 outperforms so that we take it as our encoder while adding PFA and a new branch. The last two rows suggest that estimation is more precise by using our method (Table 1).

Table 1. Experimental results for Cobb angles' estimation.

Output	Cases	MAE	MSE	SMAPE
Angle	$Vgg.$	7.78	97.58	16.6
	$Res.$	11.01	191.41	23.16
Angle + Mask	$Vgg.$	6.85	80.72	14.55
	$Vgg. + PFA$	6.69	75.65	12.97

^{a} We calculate MAE and MSE via Degree Measure.

In detail, correlation coefficients between the three different predicted angles and ground truth are given as Fig. 4. The figure demonstrates that the estimation of the first angle is much better than the rest, and the second one is better than the third one. It has the same variation tendency as angle's range that the smaller angles have worse performance (see Fig. 3).

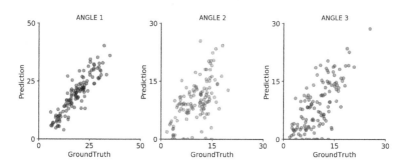

Fig. 4. The correlation coefficients between three angles predicted by the proposed method and ground truth. The angles are numbered from top to bottom and our method has much better performance on the first angle.

4   Conclusion

In this paper, we have proposed a Multi-Task network with pyramid feature aggregation to estimate Cobb angles automatically. Inspired by Multi-Task learning, we add a new branch for segmentation to enhance the feature extraction.

Typical aggregation of pyramid features is used to catch features at different levels. The two strategies both provide a more precise estimation of Cobb angles, and our method finally achieves high performance of SMAPE (12.97).

However, imbalanced estimation exists that our method estimates the top angle much better than the rest. There are several possible explanations for this result. Since big Cobb angles have a distinct appearance, the model can easily achieve excellent performance. On contrast, estimation of a straight spine with small Cobb angles has obvious relative fluctuation.

Acknowledgement. This work was supported by National Natural Science Foundation of China (Grant No. 61671399) and by the Fundamental Research Funds for the Central Universities (Grant No. 20720190012).

References

1. He, K., Gkioxari, G., Dollár, P., Girshick, R.: Mask R-CNN. In: Proceedings of the IEEE International Conference on Computer Vision, pp. 2961–2969 (2017)
2. He, K., Zhang, X., Ren, S., Sun, J.: Deep residual learning for image recognition (2015)
3. Ronneberger, O., Fischer, P., Brox, T.: U-net: convolutional networks for biomedical image segmentation. In: Navab, N., Hornegger, J., Wells, W.M., Frangi, A.F. (eds.) MICCAI 2015. LNCS, vol. 9351, pp. 234–241. Springer, Cham (2015). https://doi.org/10.1007/978-3-319-24574-4_28
4. Simonyan, K., Zisserman, A.: Very deep convolutional networks for large-scale image recognition. arXiv preprint arXiv:1409.1556 (2014)
5. Wu, H., Bailey, C., Rasoulinejad, P., Li, S.: Automatic landmark estimation for adolescent idiopathic scoliosis assessment using BoostNet. In: Descoteaux, M., Maier-Hein, L., Franz, A., Jannin, P., Collins, D.L., Duchesne, S. (eds.) MICCAI 2017. LNCS, vol. 10433, pp. 127–135. Springer, Cham (2017). https://doi.org/10.1007/978-3-319-66182-7_15
6. Wu, H., Bailey, C., Rasoulinejad, P., Li, S.: Automated comprehensive adolescent idiopathic scoliosis assessment using mvc-net. Med. Image Anal. **48**, 1–11 (2018)
7. Xue, W., Islam, A., Bhaduri, M., Li, S.: Direct multitype cardiac indices estimation via joint representation and regression learning. IEEE Trans. Med. Imaging **36**(10), 2057–2067 (2017)
8. Xue, W., Nachum, I.B., Pandey, S., Warrington, J., Leung, S., Li, S.: Direct estimation of regional wall thicknesses via residual recurrent neural network. In: Niethammer, M., et al. (eds.) IPMI 2017. LNCS, vol. 10265, pp. 505–516. Springer, Cham (2017). https://doi.org/10.1007/978-3-319-59050-9_40
9. Zhang, K., Xu, N., Yang, G., Wu, J., Fu, X.: An automated cobb angle estimation method using convolutional neural network with area limitation. In: Shen, D., et al. (eds.) MICCAI 2019. LNCS, vol. 11769, pp. 775–783. Springer, Cham (2019). https://doi.org/10.1007/978-3-030-32226-7_86

Author Index